Hidden In Plain Sight 6

Andrew Thomas studied physics in the James Clerk Maxwell Building in Edinburgh University, and received his doctorate from Swansea University in 1992.

His *Hidden In Plain Sight* series of books are science bestsellers.

Also by Andrew Thomas:

Hidden In Plain Sight
*The simple link between relativity
and quantum mechanics*

Hidden In Plain Sight 2
The equation of the universe

Hidden In Plain Sight 3
The secret of time

Hidden In Plain Sight 4
The uncertain universe

Hidden In Plain Sight 5
Atom

Hidden In Plain Sight 7
The fine-tuned universe

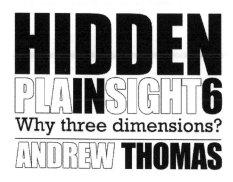

HIDDEN
PLAINSIGHT 6
Why three dimensions?
ANDREW THOMAS

**AGGRIEVED
CHIPMUNK
PUBLICATIONS**

AGGRIEVEDCHIPMUNK.WORDPRESS.COM

Hidden In Plain Sight 6

Copyright © 2016 Andrew D.H. Thomas

ISBN-13: 978-1534711914
ISBN-10: 1534711910

CONTENTS

PREFACE

Why are there only three dimensions? It is surprising just how little research there has been on why there are three dimensions of space. In fact, I believe this is only the second book dedicated to the precise question which has ever been written (the first book was written by Ptolemy, two thousand years ago).

You might imagine that the question of why there are three dimensions of space would be one of the most outstanding and pressing questions facing physics today. But it is rarely mentioned in that respect. Whenever a list is compiled of the most important questions facing physics, the usual suspects always appear: unifying gravity and quantum mechanics, dark energy and dark matter, the so-called hierarchy problem in particle physics. But you will rarely see the question arising of why there are only three dimensions. Is the question considered too difficult, and is therefore best ignored?

Admittedly, as we shall see later, string theorists might argue that there is plenty of research in string theory to provide an answer to why we only observe three dimensions – though string theory actually predicts there are ten dimensions of space! So string theory is not so much a drive to answer the question of why there are only three dimensions – it is more a drive to answer the question of why we do not observe the troublesome seven dimensions which are predicted by string theory but do not fit the bill.

This book also contains a description of the very exciting recent discovery of gravitational waves.

I have really enjoyed writing this book, probably my most enjoyable book so far. I have found it to be a fascinating topic, with lots of interesting side-issues emerging. I hope you enjoy reading it as much as I enjoyed writing it.

I think the conclusion in the final chapter is particularly convincing.

Once again, thank you for your support.

Andrew Thomas (hiddeninplainsightbook@gmail.com)
Swansea, UK,
2016

1

THE MAGIC NUMBER

In 1990, the American hip-hop trio De La Soul had one of their biggest international hit singles. The track was called *The Magic Number*, and it reached No. 7 in the UK pop chart. The song emerged at a time when rap music was dominated by violent and misogynistic themes. In contrast, De La Soul sang about love and peace and harmony, themes which were more commonly associated with the 1960s hippie community.

In the lyrics of the song, De La Soul appear quite convinced that the "magic number" is the number three. As an inquisitive physicist, I was intrigued by this apparent certainty as we clearly live in a universe which has three dimensions. An object can move in three independent directions: left/right, forward/back, and up/down. Perhaps surprisingly, we simply do not know why it is the case that there are three dimensions. Why is three genuinely a magic number in this respect?

So I was intrigued by the De La Soul track and decided to investigate the lyrics more closely. Had this talented but otherwise seemingly unremarkable trio stumbled upon a secret which had so far eluded the best minds of the scientific community?

Perhaps it was a long shot. And, after reading the lyrics, I must admit I was disappointed.

At the core of De La Soul's thesis seemed to lie the idea that three was a "magic number" purely on the basis that there were three musicians in their group. Furthermore, they appeared to regard their repeated assertion that "Three is the magic number" as some sort of substitute for a testable scientific hypothesis.

So what is so special about the number three? This opening chapter will just consider the number three. Is there anything obviously remarkable about this number? Is it inherently special?

The prime number

After my initial disappointment, I returned to consider the De La Soul track and decided that perhaps there might be some truth in what they said. I came to this conclusion because, basically, it's a great track. You don't mind that their argument is unconvincing because the lyrics work in the overall context of the song. What is more, the song was a big hit, which seems to suggest that a lot of people agreed with De La Soul: the number three certainly seemed like a plausible contender to be a magic number.

So why did the number three work in the context of the song? That question is equivalent to asking why treating the number 17, or 54, or 378.25 as the magic number simply would not work as the lyric.

"378.25, and that's the magic number." I don't think so.

Of course, an obvious answer is that the word "three" only has one syllable, whereas the phrase "three hundred and seventy eight point two five" has eleven syllables, so it simply would not fit into the rhyming structure of the line. But the word "six" has only one syllable, as does "one", "two", "three", "four", "five", "eight", "nine", "ten", or "twelve". With all these possible alternatives, why pick three as the magic number? Let us consider the properties of the number.

Firstly, three is a *natural number*, which are the numbers used for counting. For example, we might say "There are six apples on the table". The natural numbers are the non-negative integers. According to the German mathematician Leopold Kronecker: "God made the first ten numbers; the rest is the work of man." With this quote, Kronecker emphasized that the whole mighty edifice of mathematics is fundamentally based on simple numbers.

Animal bones dating back to 30,000 BC have been found with scratch marks which seem to indicate some form of early counting behaviour. This probably indicates a hunter recording his kills, or of early farmers counting livestock. It is known that shepherds frequently used pebbles to count their sheep (the word "calculate" derives from the Latin *calculus* which means "pebble"). Our natural acquaintance and use of the non-negative natural numbers for counting purposes – a direct connection with the physical world – might provide one explanation for our apparent affinity with certain natural numbers. For example, De La Soul would have been unlikely to sing "Minus five, that's the magic number".

Secondly, three is a *prime number*. It is the smallest odd prime number. A prime number cannot be divided precisely by any other number (apart from the number one). Other prime numbers include 2, 5, and 7. So clearly there are plenty of prime numbers distributed among the smaller numbers. However, the prime numbers become rarer as they get larger (because there is more likelihood of there being a smaller divisor number).

The ancient Greek mathematician Euclid proved that any natural number can be created by multiplying prime numbers together. This is called the *fundamental theorem of arithmetic*. For example, $20 = 5 \times 2 \times 2$. The reveals how the prime numbers might be considered the primordial elements from which all the numbers are constructed. Indeed, the words "prime" and "primordial" share the same root from the Greek word *primus* (meaning "first"), as does the word "primitive". This makes the prime numbers appear as if they are the fundamental building blocks of mathematics. The close link between mathematics and physics is well established. So, for this reason, we should perhaps not be surprised to find a prime number of spatial dimensions: we are dealing with one of the fundamental physical constants – the building blocks of Nature.

The fundamental nature of the primes was considered by Enrique Gracián in his book *Prime Numbers*: "In the same way that atoms combine to form molecules, prime numbers combine to form composite numbers."

Is there any reason why we might find the prime numbers intuitively more appealing? In 2014, the *Guardian* newspaper ran a poll to discover the "World's Favourite Number". After 44,000 people had voted, it was announced that the number seven was the most popular number, with the number three in second place. Both seven and three are prime numbers. Almost half the submissions were for the numbers between one and ten.

The least popular numbers were the numbers which were divisible by ten, i.e., numbers which finished with the digit zero. Perhaps those composite numbers were unpopular because they seemed rather clinical, calculated, and sterile. In contrast, the popularity of the prime numbers is possibly because they seem more fundamental, more "earthy". Indeed, in one of the quotes on the *World's Favourite Number* website (http://tinyurl.com/numberwebsite), one of the respondents said she deliberately chose the number seven because it was prime.

According to the mathematician Alex Bellos, who arranged the poll: "The point here is that we are always sensitive to arithmetical patterns, and this influences our behaviour – even if we are not conscious of it, and irrespective of our ability at maths."

So maybe that is the reason why De La Soul chose three to be their magic number.

The number of sufficient magnitude

Let us continue this examination of the role of the number three in our culture by examining how we express the number in our natural languages.

In his book *Receiving Aristotle in an Age of Crisis*, the professor of philosophy David Roochnik explores what he calls the "exceptional nature" of the number three:

> *In ordinary language, both Greek and English at least testify to the exceptional nature of the three. If my eyes hurt, and you ask me "Which one hurts?" I will answer "Both of them" rather than "All of them". The word "all" is first used when I have at least three items to count.*

Roochnik is suggesting that three is the first number to be of sufficient magnitude to be considered as representing a significant number of objects. We find similar differences in terminology when we consider the difference between the *cardinal* numbers and the *ordinal* numbers. A cardinal number is the form of a number which is used for counting, such as one, two, three, etc. The ordinal version of the number refers to a position in a ranking system, such as first, second, third, etc. We can see that the cardinal and ordinal versions of the numbers one and two are very different: "one/first", "two/second". However, the cardinal and ordinal versions of numbers greater than two are similar: "three/third", "four/fourth", "five/fifth", etc. We find this similarity of cardinal and ordinal for numbers three (or greater than three) in other languages: French, German, and Italian. So, again, it appears that once we consider the numbers three and greater, we reach a milestone in that we are now considered to have a large multiplicity of numbers. The number three is the first number which represents a significant amount, and in language we treat it differently from the numbers one and two.

The number three certainly seems to play an important role in our culture. When a group photo is being taken, the photographer will often say "I will count to three and then take the photo", which seems to reinforce the idea that three is the first number of sufficient magnitude (counting to one or two would not provide enough time for everyone to get ready).

In baseball, three is the number of strikes before the batter is out, based on the notion that three should be a sufficiently large number of attempts for any batter. Similarly, the "Three strikes and you're out" principle has also entered the American legal system. Currently, twenty four states impose harsher sentences for a third offence. As an example, California requires a criminal who has committed three serious or violent felonies to go to prison for a minimum of 25 years (and could be as much as a life sentence). The principle behind these Three Strikes laws is that three chances are deemed to be a sufficiently large number of chances to give to any criminal. Again, we find the principle of three being a number considered to be of sufficient magnitude.

In association football (soccer) a player who scores three goals is said to have scored a *hat trick*, one of the most cherished achievements. The scoring of a hat trick is celebrated much more than the scoring of two goals or four goals, and the scorer traditionally takes the match ball home as a prize. Again, the hat trick celebration reinforces the idea of three being the first number of sufficient magnitude, and the achievement of three successes being regarded as exceptional.

This book is structured rather like a detective whodunnit. We will be considering the evidence, examining various theories, rejecting some theories while retaining some theories which show promise. And – even at this early stage – I think we have found a shred of evidence here. In many aspects of life, the number three is the smallest number which is considered suitable for performing many tasks.

In the final chapter of this book it will be proposed that one and two dimensions of space would not be sufficient to perform a particular requirement of Nature – and this will emerge as a potential solution to the question of why there are three dimensions of space.

Occam's razor

William of Ockham (or "Occam") was an 14th century English monk who was one of the major figures of medieval philosophy.

Surprisingly, one aspect of William of Ockham's work has remained influential even into the modern era. His modern fame and influence is based entirely on one phrase he wrote in Latin: *Entia non sunt multiplicanda praeter necessitatem*, which means "Entities should not be multiplied beyond necessity". This phrase is commonly known as *Occam's razor*.

So why is this particularly relevant to our attempt to discover why there are only three spatial dimensions? Well, it is relevant for the simple reason that three is certainly quite a small number. Out of all the infinity of possible numbers – some of which are inconceivably huge – why do we end up with just three dimensions? We might certainly imagine that Nature has applied Occam's razor: "Entities should not be multiplied beyond necessity."

Essentially, Occam's razor is a statement in favour of simplicity. Another translation of the phrase (which is perhaps more relevant to out discussion) is: "Among competing hypotheses, the one with the fewest assumptions should be selected." Phrased in this way, Occam's razor becomes a form of guidance for physicists. It is not strictly scientific, it is not even always correct, but as a general rule it has been repeatedly shown to have value when judging the value of various hypotheses about Nature. Put simply, the simplest theory (which predicts the same results) is almost always correct.[1]

In many ways, physics could be seen as a drive towards simplicity. We seek the simplest explanations for the phenomena we observe. Simpler explanations reveal deeper underlying truth. And physics is (hopefully) a search for truth.

---

[1] I have often wondered why Occam's razor is called a "razor". Surely it does not have anything to do with shaving? Well, in the true spirit of Occam's razor, the simplest solution (and our first guess) is correct: it does indeed have something to do with shaving. Occam's razor is said to "shave" away unnecessary assumptions, cutting away the complexity.

In his book *The Fabric of Reality*, David Deutsch describes how theories which are more fundamental can be both simpler and fewer in number. Deutsch says his motivation in physics is nothing less than the goal of "understanding everything". This might seem like an impossibly ambitious goal, but, as Deutsch explains, this does not mean he wants to "know everything" – Deutsch has no interest in memorizing all the facts of the world's encyclopaedias. Instead, he wants to **understand** all the reasons why the physical world works the way it does. Understanding a comparatively simple theory (for example, general relativity) can explain a multitude of facts (the precise motion and position of the stars and planets, for example). And, while it is certainly not possible to know all the facts about the world, it might be possible to know a few simple theories which explain those facts. As Deutsch says:

Confronted with this vast and rapidly growing menu of the collected theories of the human race, one may be forgiven for doubting that an individual could so much as taste every dish in a lifetime, let alone, as might once have been possible, appreciate all known recipes. Yet explanation is a strange sort of food – a larger portion is not necessarily harder to swallow. A theory may be superseded by a new theory which explains more, and is more accurate, but it is also easier to understand, in which case the old theory becomes redundant, and we gain more understanding while needing to learn less than before.

Deutsch then presents probably the classic example of this drive toward simpler theories: the move to a *heliocentric* (Sun-centred) model of the Solar System. The ancient model developed by Ptolemy placed the Earth at the centre of the Solar System, and this *geocentric* model dominated for fifteen centuries. Because it was known that the motion of the planets was irregular (with the planets sometimes appearing to change direction) it was necessary to introduce the idea of *epicycles*, a smaller orbit contained within the planet's usual orbit around the Earth. Here we see a planet orbiting on its additional smaller epicycle:

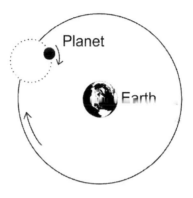

The epicycle allowed the planet to sometimes move backwards when viewed from Earth. This geocentric model of the universe was dominant until the 16th century when the Polish astronomer Nicolaus Copernicus realised that the model could be greatly simplified by placing the Sun at the centre of the Solar System and having the planets (including the Earth) orbit the Sun.

Copernicus's model was simpler than the Ptolemaic model, and it not only explained the retrograde motions of the planets but it also explained why the Earth experienced the seasons as it orbited the Sun once a year. So a simpler model can explain more.

The adoption of the Copernican model was a classic example of the value of Occam's razor in practice. Its obvious simplicity when compared to the Ptolemaic model meant it was to be preferred as an explanation.

Theories can have another quality which is similar to "simplicity" and that is "beauty". Beauty is perhaps more associated with mathematical theorems, and Bertrand Russell once described mathematical beauty in the following words:

> *Mathematics, rightly viewed, possesses not only truth, but supreme beauty — a beauty cold and austere, like that of sculpture, without appeal to any part of our weaker nature, without the gorgeous trappings of painting or music, yet sublimely pure, and capable of a stern perfection such as only the greatest art can show. The true spirit of delight, the exultation, the sense of being more than Man, which is the touchstone of the highest excellence, is to be found in mathematics as surely as poetry.*

The great physicist (and mathematician) Paul Dirac certainly believed in the importance of beauty in physics, as he wrote in 1939:

> *The research worker, in his effort to express the fundamental laws of Nature in mathematical form, should strive mainly for mathematical beauty. It often happens that the requirements of simplicity and beauty are the same, but where they clash the latter must take precedence.*

Nobel Prize-winning physicist Frank Wilczek has recently written a book called *A Beautiful Question* in which he wonders why the world is apparently so beautiful.[2] Is there any reason why that should be the case? Wilczek even

---

[2] Frank Wilczek was the joint-discover of asymptotic freedom in the strong force.

compares the world to a work of art, and his lavishly-illustrated book compares human artworks with beautiful structures from physics.

I was fortunate to meet Frank Wilczek recently. I congratulated him on his excellent book, and I asked him whether "beauty" was rather a subjective concept. In reply, he felt that there were clear objective standards of beauty, which had remained consistent down the ages: everyone seems to agree on "great art". He had gained a very strong appreciation for renaissance art in particular. He felt that the discovery of perspective by renaissance artists was an example of objectivity in art: "By understanding how the same scene can appear different, depending on the viewpoint from which it is perceived, we learn to separate the accidents of viewpoint from the properties of the thing itself. By treating subjectivity objectively, we master it."

Here is a photo of me with Frank Wilczek:

According to Frank Wilczek, Nature's "artistic style" has two obsessions:

- Symmetry – a love of harmony, balance, and proportion.

- Economy – satisfaction in producing an abundance of effects from very limited means.

The importance of symmetry – and especially balance – in fundamental physics was explored in my previous book. The importance of economy – "producing an abundance of effects from very limited means" – reflects the influence of Occam's Razor.

Beauty, however, still seems to me to be rather a subjective notion to evaluate possible theories: what you might find to be a beautiful theory, I might find ugly. However, that is not the case for simplicity, because, surprisingly, it is possible to obtain a quantitative measure of the simplicity of a theory.

A measure of simplicity can be found because simplicity is the opposite of *complexity*, and there is a large body of science which deals with complexity.

How can we measure complexity (how complicated something is)? Well, consider a pattern, say, a square of black and white dots, or maybe a binary sequence or zeroes and ones. How complex would you say the pattern is? How would you measure the complexity of the pattern? Or, imagine you have two patterns, two squares of black and white dots. Which pattern would you say is the most complex? How could you decide between the two patterns?

Well, it turns out that we can measure the complexity of the patterns quite objectively. There is a measure called *Kolmogorov complexity*, which sounds quite complex in itself but is actually quite a simple and clever idea. The Kolmogorov complexity of a pattern can be found on the basis that a computer program could be used to calculate the pattern, and print out the black and white dots. As an example, if a pattern of zeroes and ones is a repeating pattern then the computer program to print the numbers would be quite short: just a few lines, with the code probably featuring a loop. For example, the repeating pattern:

xyxyxyxyxyxyxyxyxyxyxyxyxyxyxyxy

which has 32 characters could be represented by the code "xy 16 times", which only has 11 characters. This shows it is possible to compress the pattern to a smaller length, while still retaining the same amount of information.

Alternatively, if the pattern of zeroes and ones is truly a complex pattern, then the computer program needed to print the pattern could be very long. In the extreme case – for the most complicated patterns – the computer program would be as long as the pattern. In that case, it would not be possible to make the computer program any shorter than the original pattern. The minimum length of the computer program needed to print the pattern is called the Kolmogorov complexity of the pattern. We can see that the Kolmogorov complexity describes the *compressibility* of the pattern: if the pattern can be produced by a computer program which is shorter than the actual pattern, then we can say that the pattern can be compressed to a shorter length (it can be represented by a computer program with a shorter length).

So we see we can obtain an objective measure of the complexity – or, inversely, the simplicity – of any pattern. And that is very useful for our purposes because a physics theory is, after all, nothing more than a pattern of letters when it is written down. So we can objectively measure the simplicity of any theory. This idea is expressed by the mathematician Gregory Chaitin:

*This idea of program-size complexity is also connected with the philosophy of the scientific method. You've heard of Occam's razor, of the idea that the simplest theory is best? Well, what's a theory? It's a computer program for predicting observations. And the idea that the simplest theory is best translates into saying that a **concise** computer program is the best theory. What if there is no concise theory, what if the most concise program or the best theory for reproducing a given set of experimental data is **the same size** as the data? Then the theory is no good, it's cooked up, and the data is incomprehensible, it's random. In that case the theory isn't doing a useful job. A theory is good to the extent that it compresses the data into a much smaller set of theoretical assumptions. The greater the compression, the better!*

So whereas it is impossible to objectively measure the **beauty** of a physics theory (your idea of beautiful might be my idea of ugly), it is possible to objectively measure the **simplicity** of a theory.

It therefore appears to be a valid conjecture that Nature is described by the simplest set of rules, and we should therefore apply Occam's razor in an attempt to minimise the simplicity of our theories in order to obtain the most accurate model of the workings of Nature.

The natural number

There are a many important constants (special numbers) in mathematics, some of which are quite famous. Perhaps the most famous of the mathematical constants is pi (π) which is the ratio of a circle's circumference to its diameter, and has the approximate value 3.14159. Other important mathematical constants include Euler's number (e) which has the approximate value 2.718, and the golden ratio which has the approximate value 1.618. For a full list of these constants, refer to the Wikipedia page on "Mathematical constants". [3]

If you consider the list of mathematical constants on the Wikipedia page, you might discover something surprising: almost all of the constants have a value quite close to one. The largest constant listed is the Feigenbaum constant which has an approximate value of 4.669. So none of the listed constants have large values of the order of 10,000, or several million. This is a known fact about the mathematical constants: they tend to be close to one. This is a very useful fact as far as physicists are concerned.

These mathematical constants can often be produced by an infinite series of calculations, each calculation producing a more accurate result. For example, in 250 BC the Greek mathematician Archimedes realised that pi could be approximated by fitting a regular polygon inside a circle, and then adding up the lengths of all the sides of the polygon.

---

[3] https://en.wikipedia.org/wiki/Mathematical_constant

The more sides the polygon has, the closer to pi the approximation becomes:

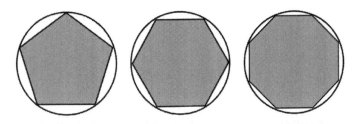

So why are the values of these mathematical constants close to one? Well, if you consider the form of these infinite series used to calculate the mathematical constants, you will find that the first term in many of these series is the number one. For example, one infinite series which can be used to calculate pi is:

$$\frac{\pi}{4} = 1 - \frac{1}{3} + \frac{1}{5} - \frac{1}{7} + \frac{1}{9} \dots$$

and an infinite series to calculate Euler's number, e, is:

$$e = 1 + \frac{1}{1!} + \frac{1}{2!} + \frac{1}{3!} + \frac{1}{4!} \dots$$

and an infinite series to calculate the golden ratio, φ, is:

$$\varphi = 1 + \cfrac{1}{1 + \cfrac{1}{1 + \cfrac{1}{1 + \dots}}}$$

You will see that the first term in all of these infinite series is the number one (with the following terms becoming progressively smaller). So we can see how the mathematical constants are frequently derived from the number one, and so tend to have low values. This was described by Einstein in a letter to an old student friend, Ilse Rosenthal-Schneider (you will note how he refers to the calculation of Euler's number, e, using the infinite series we considered earlier):

I see from your letter that you did not grasp my hint about the universal constants of physics. I will therefore try to make the matter clearer.

Basic numbers are those which, in the logical development of mathematics, appear by a certain necessity as unique individual formations. For example:

$$e = 1 + \frac{1}{1!} + \frac{1}{2!} + \frac{1}{3!} + \frac{1}{4!} \cdots$$

It is the same with π, which is closely connected with e. In contrast to such basic numbers are the remaining numbers which are not derived from 1 by means of a perspicuous construction.

It would seem to lie in the nature of things that such basic numbers do not differ from the number 1 in respect of the order of magnitude, at least as long as consideration is confined to "simple" or, as the case may be, "natural" formations.

There are constants in physics as well, for example, the speed of light, or the gravitational constant. It is clear that these values can quite possibly have extremely huge (or small) numerical values depending on the units selected to express the constant. For example, the speed of light is the extremely large value of 3×10^8 m/sec. However, if we choose a different unit – for example, using kilometres as the unit of length instead of metres – then we get a smaller number: 3×10^5 km/sec (note the change of unit). So by choosing different units, we change the numerical value of the physical constant.

However, there are some constants in physics which are just plain numbers, i.e., they have no associated units. One of the most famous is the *fine structure constant*, which is related to the strength of the electric charge and has the numerical value of 1/137 (note: in this case there are no units involved – it is just a number). So these values always stay the same – irrespective of units (because they have no units). Even a civilisation of aliens in a different galaxy would calculate exactly the same numerical value for these constants. That makes them particularly interesting for a physicist as they seem to reflect very deep truths about Nature.

Because mathematical constants tend to be close to one, there is justification to expect the physical constants (which – at the lowest level – are surely derived from the mathematical constants in some way, via some mechanism) to also be close to one. (Remember: this is only true for physical constants which are not expressed in any particular units).

So we are entitled to expect any physical constants (which are not expressed in units) to be close to the number one. This principle is called *naturalness* (and the physical constants are then said to be *natural*).

This insight into the importance of naturalness is generally ascribed to Einstein, and it was described in his only publication on the subject in the German journal *Annalen Der Physik* in 1911:

> *The unit-free numerical factors, whose magnitude is only given by a more or less detailed mathematical theory, are usually of the order of unity. We cannot require this rigorously for why shouldn't a numerical factor like $(12\pi)^3$ appear in a mathematical-physical deduction? But without doubt such cases are rarities.*

This extremely fortuitous principle has proven to be a useful tool for physicists. As John Barrow explains in his book *The Constants of Nature*: "In every formula we use to describe the physical world, a numerical factor appears ... which is almost always fairly close in value to 1 and they can be neglected, or approximated by 1, if one is just interested in getting a fairly good estimate of the result."

Conversely, if we encounter any physical constants which are very far from the value of one, then we should suspect that we are not at the fundamental level, and we are missing some deeper mechanism. Our formula is not describing a fundamental process. There is some additional fundamental behaviour which needs explaining because it is not being captured by our formula. We need to dig deeper.

And, of course, this entire book is based on one particular unit-free constant, and that is the number of dimensions of space. The number three, used merely for counting in this way, is just a number. It is not three metres, or three seconds, or three kilograms. It has no units. It is just three. It is just a number. And, on the basis of Einstein's insight, this is precisely what we should expect to find. Three is a **natural** result for the number of dimensions. It is natural because it is a number close to one. The closer to one, the better, the more in line with our expectations.

So, by arguments purely based on naturalness, we should not expect there to be 10,000 dimensions of space. We should not expect there to be 100 dimensions of space. It might be argued that we should not even expect to find ten dimensions of space.

However, even without knowing the precise underlying mechanism which generates the dimensions, we would expect there to be (approximately) three dimensions. Certainly, we would expect a number close to one. Our task is then to discover the fundamental mechanism which naturally produces that result.

The Standard Model of particle physics is our best model of the physical world and has been tested to remarkable accuracy. However, there remains approximately 20 unit-free constants in the Standard Model which appear to have arbitrary values and have to be set on the basis of experimental ("empirical") measurement. Einstein believed that the presence of these arbitrary constants reveals our theory is not truly fundamental:

> *In a reasonable theory, there are no unit-free numbers whose values are only empirically determinable. Of course, I cannot prove this. But I cannot imagine a unified and reasonable theory which explicitly contains a number under which the whim of the Creator might just as well have chosen differently, whereby a qualitatively different lawfulness of the world would have resulted.*

So, how can these arbitrary constants be eliminated, thus creating a theory which is more fundamental? This is possible if we strive to create more natural theories. If all arbitrary constants in a theory have a value which is close to one, then those arbitrary constants effectively disappear (because multiplying by one is equivalent to not multiplying at all: the original term remains unchanged). So a natural theory is a simpler theory: it would have fewer arbitrary constants. We can therefore think of naturalness as arising from Occam's razor: the simplest theories are more natural.[4]

Occam's razor and the principle of naturalness are therefore closely related, and it appears that both act to drive down the number of spatial dimensions to a low value. Working in the opposite direction, we find the principle that three is the first "number of sufficient magnitude" to perform a certain task – which acts to drive up the number of spatial dimensions (as shown in the following diagram).

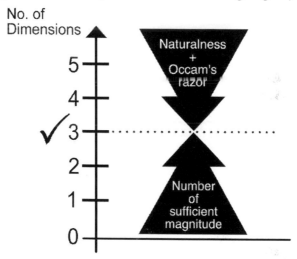

On that basis, our quest to discover why there are three dimensions becomes a quest to discover why one and two dimensions of space would not be sufficient to perform a particular requirement of Nature. We will be returning to this theme in the final chapter of the book.

---

[4] I believe this apparent connection between naturalness and simplicity is the reason for the current popularity of naturalness as a guiding principle in developing new theories – even though it might not always be applicable to all areas, such as particle physics. See Sabine Hossenfelder for a discussion: http://tinyurl.com/sabinenaturalness

So is three really a magic number, as De La Soul suggested? Well, it's starting to look like, yes, De La Soul might have been onto something after all.

2

DIMENSIONS

When we say there are three dimensions of space, what exactly do we mean? What is a dimension? This question will be answered for the specific case of the spatial dimensions, but then a more general concept of a "dimension" will be introduced.

Put simply, the three spatial dimensions refer to the usual concepts of depth, width, and height. We can specify a position in three-dimensional space using three coordinates:

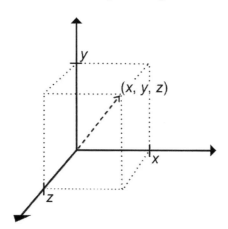

In the previous diagram, you will see that the dashed arrow points to a point in space which is a distance x along the horizontal axis, and is a distance y along the vertical axis, and a distance z along the axis which is coming out of the page. We can therefore define the position of the point by a list of its coordinates: (x, y, z).

It is possible to consider spaces which do not have three dimensions. For example, position on a flat plane (such as graph paper) could be described by just two coordinates: x and y. Hence, a flat plane would represent a two-dimensional space. Position on a line could be represented by just a single coordinate: the length along the line. Hence, a line represents a one-dimensional space. We can see that the number of coordinates required to define a point determines the number of dimensions.

No matter what the number of dimensions, it is important to be able to measure the distance between two points. A formula which can be used to calculate distances is called a *metric*. Let us first consider space with just two dimensions, described by two coordinates: x and y. In that case, the following diagram shows that travelling from point A to point B necessitates travelling a distance x horizontally, and a distance y vertically:

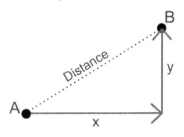

This forms a right-angled triangle, so from Pythagoras's theorem we could calculate this distance as the square root of x^2 plus y^2. So this gives us our metric on a flat two-dimensional plane:

$$d = \sqrt{x^2 + y^2}$$

As shown in the earlier diagram of three-dimensional axes, it is easy to extend this metric to three-dimensional space just by adding the extra z coordinate:

$$d = \sqrt{x^2 + y^2 + z^2}$$

Even though we cannot visualise spaces with more than three spatial dimensions, it is very easy to describe them mathematically: we simply add extra coordinates to our list which describes a point. For example, a point in a four-dimensional space might be described by (w, x, y, z). Once, again, the number of coordinates determines the number of spatial dimensions.

The metric for that four-dimensional space is also easily generated by adding another coordinate:

$$d = \sqrt{w^2 + x^2 + y^2 + z^2}$$

OK, so that is a basic description of the concepts involved when we describe spatial dimensions. But we have not yet considered the crucial question ...

What is a dimension?

We all have an instinctive understanding of space which allows us to move freely, avoiding obstacles, and performing everyday tasks such as driving a car. Unfortunately, this very intimate connection we have with space – acquired over millions of years of evolution – makes it very difficult for us to stand back and be objective about the nature of space. We all live inside space – we cannot observe it from outside. So we find it very difficult to be objective when answering the question "What is a spatial dimension?"

However, one of the most important skills of being a good physicist is to be able to ignore our human preconceptions and intuitions, and to have an open analytical mind, able to consider the data in an objective manner. And when we apply that objectivity to the subject of space and dimensions, we realise that the spatial coordinates of an object, (x, y, z), just represent properties of that object, just as the colour of an object is another example of the property of an object. Each coordinate represents a dimension of space; each coordinate represents a way in which an object's properties can be modified. We may interpret variation of those spatial coordinates as "an object moving around in space", but that interpretation gains us nothing more than just considering the variation in the positional property of an object. If an object moves, its positional property varies. Likewise, if an object changes colour, its colour property varies.

I particularly like the following quote from the German philosopher Hans Reichenbach which makes it clear how all an object's properties can be treated as dimensions, with no distinction between space and colour:

Let us assume that the three dimensions of space are visualized in the customary fashion, and let us substitute a colour for the fourth dimension. Every physical object is liable to changes in colour as well as position. An object might, for example, be capable of going through all shades from red through violet to blue. A physical interaction between any two bodies is possible only if they are close to each other in space as well as in colour. Bodies of different colours would penetrate each other without interference. If we lock a

28

number of flies into a red glass globe, they may yet escape: they may change their colour to blue and then be able to penetrate the red globe.

So, if we can treat a dimension as a range of property values, then we can also treat a range of property values as a dimension. For example, let us continue this theme of treating colour as a property which can be expressed as a dimension. It is the case that any colour can be generated by mixing together various amounts of red, green, and blue primary colours (this is how a computer monitor can generate any colour: each point on the screen is composed of microscopic red, green, and blue dots). So we can represent any single colour – yellow or purple, for example – as particular combinations of red, green, and blue. Hence, any colour can be represented as a point in a three-dimensional space, with the three axes of that space representing the three primary colours. As Frank Wilczek says in his book *A Beautiful Question*: "We can specify any perceived colour by saying how much red, how much green, and how much blue it takes to match it. This is completely analogous to how we can specify a place in space by saying how far it is in the north-south, east-west, and vertical directions. Ordinary space is a three-dimensional continuum, *and so is the space of perceived colours*."

Continuing this theme of treating property values as a dimension, Lisa Randall gives several examples in her book *Warped Passages*: "When you peg someone down as one-dimensional, you actually have something rather specific in mind: you mean the person has only a single interest." Lisa Randall presented the example of Sam, who sits at home all day watching sports on TV. The number of hours Sam watches TV could be represented as a point in a one-dimensional space.

Let us consider a similar example. Let us consider a book with a certain number of pages. Again, the book can be represented as a point in a one-dimensional space:

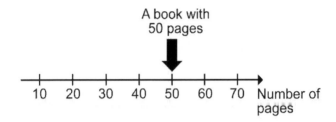

Let us now consider a second book-related property: the colour of a book's cover. This can be represented as a point in a continuous spectrum of all possible colours. The following diagram shows a book with a green cover represented by a point in a different one-dimensional space:

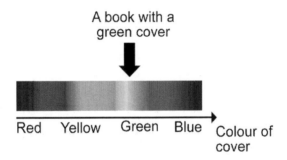

There is clearly a unifying theme behind these two one-dimensional properties: they can both refer to a single book. So let us combine these two properties, forming a two-dimensional space:

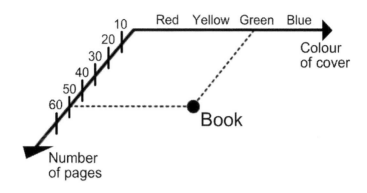

From the diagram above, it can be seen how a book can now be completely described by a single point in a two-dimensional space (instead of previously being described by two separate points in two different one-dimensional spaces). From that single point in the diagram above, you can see that we can determine that the book has fifty pages, and has a green cover.

So it is possible to unify two different descriptions by creating a higher-dimensional space, and we shall see later in this book how this method of unification has become very popular in physics.

Considering the book example, this is all well and good: the unification of page count and cover colour was based on a single underlying entity (a book), and so the unification made good logical sense. However, it is very easy to apply this technique to any two different descriptions – even if they have nothing in common! Mathematically, we can combine any two descriptions into a single higher-dimensional space, and the result will be valid. But, in physics, the big question is whether it is legitimate to perform the unification in this way. Does the unification represent a true underlying connection? Does there exist an object which logically combines the two descriptions? Or are we artificially "glueing" the two descriptions together, with no actual basis in reality?

As an example, we might count the number of wings on an animal (for example, a bird has two wings, but a bumblebee has four wings), and we might count the number of tentacles on an animal (for example, an octopus has eight tentacles, but a jellyfish can have more than a hundred), and represent these two descriptions (wings and tentacles) as two points in two different one-dimensional spaces. However, we could then "glue" these two descriptions together to form a single point in a two-dimensional space. That single point would tell us the number of wings on a particular animal, **and** the number of tentacles on that animal – we have unified two apparently completely different descriptions!

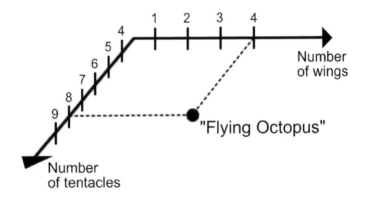

That might appear to be a great achievement, however, this is a totally artificial approach as the combined two-dimensional object – a "flying octopus" – does not exist in Nature.

As we shall see later, this unification approach in physics using higher dimensions has become very popular. As we have just seen, it is an easy way of achieving an impressive result – almost too easy. Basically, we can unify any two properties using higher dimensions. But we have to ask ourselves, while the resultant unification might be valid mathematically, does it have physical relevance?

Or are we just making flying octopuses?

Spacetime

One example of this type of unification is the "glueing" together of three-dimensional space and one-dimensional time in order to produce a higher-dimensional four-dimensional *spacetime*. It was actually Einstein's former mathematics tutor at the Zurich Polytechnic – Hermann Minkowski – who realised that it made sense to combine space and time to form spacetime. Minkowski realised that his former pupil's theory of special relativity indicated that time and space were not independent entities, but in fact motion through space could affect motion through time (famously, the astronaut who flies away from Earth at a speed close to the speed of light will find – when he returns to Earth – that he has aged less than someone who remained on Earth).

As Hermann Minkowski said:

Henceforth space by itself, and time by itself, are doomed to fade away into mere shadows, and only a kind of union of the two will preserve an independent reality.

The "glueing" together of the space and time dimensions to form a four-dimensional spacetime is valid because there exists a physical entity which must be described by **both** a point in space **and** a point in time. That physical entity is the *event*. We are used to dealing with events in everyday life: a public meeting might be advertised at occurring at a certain venue at a certain time. The meeting is therefore described by an event incorporating space and time. But even at the level of fundamental particle physics we have to deal with events: the decay of a particle can only be described by both the position of the particle and the time at which the particle decay occurred. So the concept of the event has genuine physical significance. Hence, we are most certainly not creating a "flying octopus" – a unification with no physical significance – when we glue together space and time.

Absolute space

Let us now consider a concept which was introduced by Isaac Newton in the *Principia* which has implications for our quest to determine why there are three dimensions of space (or, equivalently, four dimensions of spacetime). As described in my first book, Newton was a firm believer in *absolute space*, the idea that space took the form of a permanent grid-like background inside which objects could move:

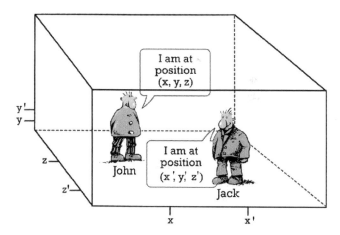

Essentially, absolute space represents a three-dimensional "box". Even if all the matter in the entire physical universe was removed, Newton believed absolute space would still exist as an entity in its own right. This certainly has consequences for our quest into the origin of three dimensions: is there a pre-existing three-dimensional "box"?

However, Newton's idea of absolute space has been categorically refuted. Instead, it is now accepted that individual observers – moving at constant velocity – can define their own *inertial frame of reference*, essentially defining their own set of coordinate axes. And the principle of relativity tells us that the laws of physics are the same for all observers – regardless of their velocity. Hence, every observer's frame of reference is equally valid: there can be no single definition of absolute space which can apply to all observers. An equivalent argument is that absolute space could be used to determine if an observer was absolutely stationary, but relativity tells us that there is no experiment which can distinguish between being stationary and moving at constant velocity.

Instead of absolute space, Newton's great rival Gottfried Leibniz proposed that space must be relative, with spatial positions emerging from the relationships between the objects within the universe:

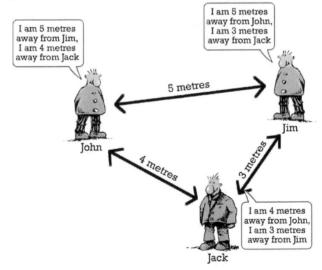

In the model of relative space, space emerges from the properties of the objects **within** the universe. I believe this is an important clue in our quest to discover why there are three dimensions of space: we should expect three-dimensional space to **emerge** from the properties of the objects (fundamental particles) which compose the universe, rather than expecting the dimensions to already exist as some absolute background three-dimensional axes.

We will be returning to this theme in the final chapter.

3

THE GEOMETRY OF THREE DIMENSIONS

In this chapter, we will be considering the geometry of three-dimensional space. Is there anything particularly special about three-dimensional geometry? Does it possess any particular properties which are not shared by geometries in other numbers of dimensions?

Let us start by considering *polygons*. A polygon is simply a shape composed of a connected sequence of straight lines forming a loop. A rectangle is an example of a polygon with four sides. The simplest polygon which can exist is the triangle with three sides (again: the magic number). Bearing this in mind, perhaps the number three is the smallest allowable number which can represent the existence of a physical body? A "number of sufficient magnitude"? The triangle is also the most stable physical shape which is widely used in construction and design, perhaps most notably in the *geodesic dome*. The geodesic dome is the most stable structure which can be produced from the least amount of materials.

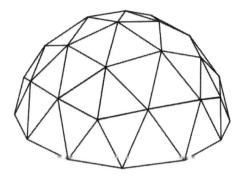

Does Nature take advantage of the strength of this triangle-based design principle? We will find out later.

Of particular interest are the *regular polygons*. These are polygons which have all sides equal, and all angles equal. A square is the regular polygon with four sides, and the equilateral triangle is the regular polygon with three sides.

The following diagram shows the regular polygons with three, four, five, and six sides (including the regular pentagon and the regular hexagon).

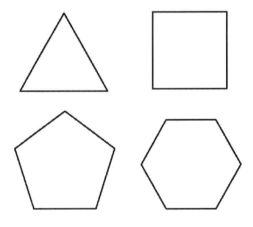

Interestingly, there are only three possible ways to tile a plane using regular polygons. The following diagram shows how this can only be achieved by using equilateral triangles, squares, and regular hexagons. Any other type of regular polygon (for example, a regular pentagon or regular octagon) cannot tile the plane.

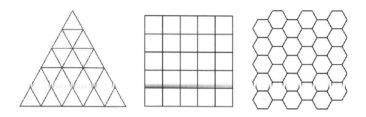

So there is that magic number again: the number three.

The Platonic solids

Let us now move from two dimensions into three dimensions. A three-dimensional object which is built from connected polygons is called a *polyhedron* (plural: *polyhedra*). As an example of a polyhedron, a cube can be constructed from six squares, one square on each face. A polyhedron which is constructed purely from regular polygons is called a *regular polyhedron*. Only five regular polyhedra exist in three-dimensional space. These are the tetrahedron, the cube, the octahedron, the dodecahedron, and the icosahedron. These five regular polyhedra are shown in the following diagram:

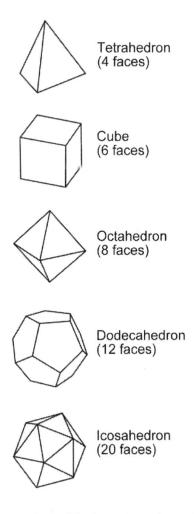

Tetrahedron
(4 faces)

Cube
(6 faces)

Octahedron
(8 faces)

Dodecahedron
(12 faces)

Icosahedron
(20 faces)

These five regular polyhedra – the only regular polyhedra which exist – have been known since antiquity. They are more commonly known as the *Platonic solids*, named after the ancient Greek philosopher Plato. Plato considered these objects to be perfect forms, inhabiting a special realm of mathematical perfection. In contrast, all physical objects on

Earth – such as balls and rough cubes – could only ever be approximations of these perfect mathematical forms.

Interestingly, Platonic solids appear in the natural world. As an example, the shape of *radiolaria* – single-celled marine life – can take the form of a regular icosahedron:

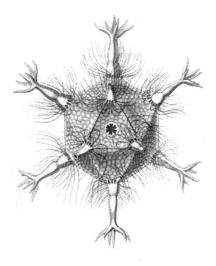

So, in radiolaria, we find an example of Nature taking advantage of the strength of a triangle-based design, just like the geodesic dome we considered earlier. In his book *A Beautiful Question*, Frank Wilczek considers the design of the herpes virus, which is another Platonic solid. He suggests that this is an example of the simplicity of Occam's razor:

> *This is a case of simplicity giving the appearance of sophistication, or more precisely of simple rules giving rise to apparently complex structures that on reflection become ideally simple. The point is that the DNA of viruses, which must instruct them in all facets of their existence, is very limited in size. To economise on the length of the construction manual, it helps to make*

your product from simple, identical parts, identically assembled. Because the part generates the whole, the virus does not need to know about dodecahedra, or icosahedra – but only about triangles, and a rule or two for latching them together.

Because all the faces of a Platonic solid are exactly the same, they are often used to make dice. As each face is identical, there is an equal chance of each face being selected. Hence, the dice are perfectly fair. You have surely seen a cube used to make dice, but the other four Platonic solids are also used to make dice for role-playing games:

As there are only five Platonic solids, there can only ever be these five fair dice.

In the 16th century, the German astronomer Johannes Kepler created a model of the Solar System in which the orbits of each of the then-known five planets related to one of the five Platonic solids. In Kepler's model, the Platonic solids were placed inside each other, and the outer edges of the solids was supposed to provide the correct orbital distance of the related planet. Again, this is based on Plato's idea of the Platonic solids representing some ideal form. The following diagram shows Kepler's model:

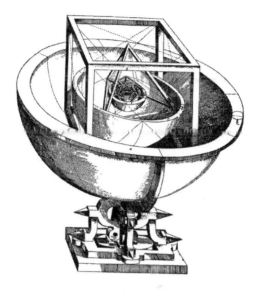

Kepler abandoned this idea as it was not accurate enough, but he did later calculate the formula for the correct motion of the planets, which was based on elliptical orbits.

The Platonic solids in higher dimensions

So far, we have considered geometry in two and three dimensions. However, as discussed in the previous chapter, we can consider the geometry of any number of dimensions. Of particular interest to us is the construction of Platonic solids in spaces with more than three dimensions. It is impossible to visualise such structures, but their construction follows the same rules as the construction of the Platonic solids in three dimensions: every side has to be a regular polygon, and every angle has to be the same.

What is particularly interesting is the number of Platonic solids which can exist in spaces with different numbers of dimensions. We have seen that only five Platonic solids exist in three-dimensional space. Perhaps it might be imagined that spaces with higher numbers of dimensions would permit greater numbers of Platonic solids to exist, but this is not the case.

We have just considered the three-dimensional case, with only five regular polyhedra – the Platonic solids – being possible in three dimensions.

As we move to consider greater than three dimensions, the polyhedra are now given the name *polytopes*. In four dimensions, it turns out that there are six regular polytopes. This is perhaps not very surprising: we have added another dimension, and discovered one extra regular polytope.

However, as soon as we go beyond four dimensions, the sequence changes completely. In five dimensions, there are only three possible regular polytopes. In six dimensions, there are – again – only three possible regular polytopes. In fact, for every number of dimensions greater than four, there are only three possible regular polytopes.

The mathematical physicist John Baez has considered the possible implications of this remarkable fact on his website:[5]

> *You might think things would keep getting more complicated in higher dimensions. But it doesn't! Four-dimensional space is the peak of complexity as far as regular polytopes go. From then on, it gets pretty boring. This is one of many examples of how four-dimensional geometry and topology are more complicated, in certain ways, than geometry and topology in higher dimensions. And the spacetime we live in just happens to be four-dimensional. Hmm.*

In that last sentence you can see that John Baez clearly considers it possible that this result has some significance as to why there are four spacetime dimensions. I also think this is a significant result, but only in a general sense. As John Baez says, it shows that the geometry of three and four dimensions has particularly interesting properties which are not shared by the higher dimensions. Cambridge University physicist John Barrow has published a comprehensive review paper entitled *Dimensionality* (a link is in the footnote[6]). In *Dimensionality*, Barrow examined the mathematical properties exhibited by the lower numbers of dimensions, "examining whether or not 3 and (3+1) dimensions lead to special results in pure mathematics. Remarkably, it does appear that low

[5] John Baez, *Platonic Solids in All Dimensions*, http://tinyurl.com/johnbaez

[6] J.D. Barrow, *Dimensionality*, http://tinyurl.com/barrowdimensionality

dimensional groups and manifolds do have anomalous properties."

We will be returning to consider the special geometrical properties of three-dimensional space in the final chapter, where we will be uncovering what I believe to be the truly important unique property possessed by three-dimensional space, and the reason why space must have three dimensions.

The stability of orbits

It is surprising that the three-dimensional geometry of space plays a central role in many equations in physics, in ways that are not immediately obvious. As an example, we can consider Newton's formula for the attractive gravitational force, F, between two masses, m_1 and m_2:

$$F = \frac{Gm_1m_2}{r^2}$$

where G is the gravitational constant.

You will see that the distance between the two masses, r, is on the bottom line of the fraction. This means there is an inverse relationship between the strength of the gravitational pull between the two objects, and the distance between the two objects: if the objects are closer (the distance is smaller) then the force of attraction is larger, and if the objects are further away (the distance is larger) then the attraction is smaller. But what is particularly interesting is that in the equation, the distance r is squared.

In her book *Warped Passages*, Lisa Randall describes this effect: "It is known as an *inverse square law*, which means that the strength of gravity decreases with distance proportionally to the distance squared. For example, if you double the distance between two objects, the strength of their gravitational attraction goes down by a factor of four."

Why is the distance squared? In the case of inverse square laws we can see a very clear logic determining why these formulas have that form, and it is due to the three-dimensionality of space.

We can imagine a mass as being a "source" of a gravitational field, and a second mass will be attracted to that first mass of the basis of the strength of that field in the immediate locality of the second mass. We can also imagine a spherical region of space, with the source mass at the centre of that spherical region:

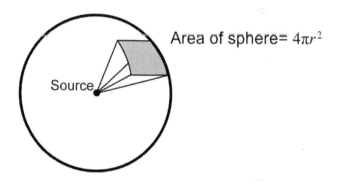

Area of sphere= $4\pi r^2$

The gravitational attraction from the source mass is clearly going to be distributed evenly around the spherical region. As we move further away from the source mass, we can see that the spherical region will grow larger (the sphere will be larger). Hence, the influence of the source mass will become more and more diluted the further the distance from the source mass. So this is the true reason why gravity becomes increasingly weaker with distance.

The formula for the area of a sphere is $4\pi r^2$, where r is the radius of the sphere. So the sphere area – and hence the rate at which gravity weakens – is proportional to the square of the distance from the source mass (the square of the radius of the sphere). So now we see the true origin of the inverse square law of the force of gravity: it is caused by the dilution of the effect of the source as it radiates into three-dimensional space. And because of the purely geometrical nature of this effect, the inverse square law also applies to many other phenomena such as the weakening of an electric field, or light, or sound.

As Lisa Randall says in *Warped Dimensions*: "The way in which the gravitational force law depends on distance, which is encoded in Newton's inverse square law, is intimately connected to the number of spatial dimensions. This is because the number of dimensions determines how quickly gravity diffuses as it spreads out into space."

But what if there are more than three dimensions of space? For example, what if there were four dimensions of space? In that case, gravity would no longer be described by an inverse square law, instead it would be described by an inverse cube law. This would result in the force of gravity varying with distance at a greater rate: in four dimensions, gravity would become very much stronger if objects moved closer together, and would become very much weaker if objects moved further apart. This effect would become even more pronounced if there were more than four spatial dimensions.[7]

---

[7] So here we find a reason for the mysterious naturalness of the physical constants (discussed in Chapter One). The value of the numerical constants in our formulas is heavily-dependent on the dimensionality of space (see Chapter Ten of John Barrow's book *The Constants of Nature*). So because the number of dimensions of space is natural (three is a small dimensionless constant) this ensures that the other constants are also natural: the physical constants are natural because space is natural.

We will now consider the important consequences of this fact for the stability of the planetary orbits (such as the orbit of the Earth around the Sun).

It was Isaac Newton who first realised that objects could be placed into stable gravitational orbits. Newton came to this conclusion by considering an ingenious thought experiment which is now known as *Newton's cannonball*.

Newton imagined an enormous cannon on top of a very high mountain. The cannon is aimed in a direction which is perfectly parallel to the Earth's surface (see the following diagram). Newton realised that if the cannon fired a cannonball at a low speed then it would reach the surface of the Earth after having travelled just a few miles, following trajectory A in the following diagram. Newton also realised that if the cannonball was fired at a tremendously high speed then it would travel straight into space along trajectory C. Newton's great insight was that there had to be an intermediate speed between those two velocities which would result in the cannonball following trajectory B in the diagram below:

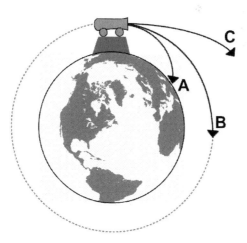

You can see from the diagram that if the cannonball follows trajectory B it will travel on a course which will take it completely around the Earth back to its starting point. Hence, that cannonball could be placed into a perpetual stable orbit around the Earth.

The speed at which the cannonball (or any object) must be fired to remain in that orbital path is called the *orbital velocity*.

Let us now consider the forces involved when an object is in orbit. Let us consider a satellite – for example, the International Space Station – in orbit around the Earth. I suspect it is generally believed that satellites such as the International Space Station stay in orbit in space because gravity is virtually non-existent in space, so satellites are weightless. This is most certainly not the case. The International Space Station is in orbit only 250 miles above the surface of the Earth. I live in the west of Britain, which is not a large country, but I am frequently nearer the International Space Station than I am to London. At that low height, the force of gravity is only slightly weaker than it is on the Earth's surface (the strength is actually 88% of the strength of gravity on the surface of the Earth). If you dropped an object from that height, it would most certainly fall toward the Earth at great speed.

No, the reason the International Space Station (and all other satellites) stay in orbit is because they are travelling so fast – as anyone who has seen the International Space Station shooting across the night sky will agree.[8] The

---

[8] It is quite easy to see the International Space Station at night: it is the third brightest object in the sky and it is amazing to see it passing overhead at great speed. There is a website which tells you when you

International Space Station is travelling at 17,100 mph. As you know, if you have a heavy bucket attached to a rope, and you swing the bucket around yourself, it will feel as if the bucket is pulling on the rope. This apparent outward force is called *centrifugal force*.[9] It is this force which holds the satellite in orbit above the surface of the Earth.

The formula for the value of the centrifugal force, F, is:

$$F = \frac{mv^2}{r}$$

where m is the mass of the orbiting object, v is the velocity of the object, and r is the orbital radius.

Now let us consider the stability of an orbit. What would happen if the International Space Station slowed slightly, maybe after being hit by a rock? This would result in a considerable reduction of centrifugal force which, as we see from the formula, is proportional to the square of the velocity of the satellite. Gravity remains as strong as ever, so would this reduction in centrifugal force result in the satellite rapidly plummeting to the Earth, in an ever-accelerating descent? In other words, is the orbit unstable?

Well, the quick answer is no. And this is because of the law of conservation of angular momentum. The formula for angular momentum, L, is given by:

---

can see the station: http://iss.astroviewer.net/observation.php

[9] The tendency of the bucket is to fly off in a straight line. So the only force actually acting on the bucket is the tension in the rope (this inward force is called the *centripetal force*). So the apparent outward centrifugal force you feel is considered to be a *fictitious force*.

$$L = mvr$$

so:

$$v = \frac{L}{mr}$$

This formula shows that as the radius of the orbit gets smaller, the velocity increases (given that angular momentum is conserved: the value of the L term stays constant). This is the same principle as an ice skater spinning on the spot, who brings their arms in closer together in order to spin faster.

The smaller radius brings greater speed, and it is this greater speed which saves the International Space Station from plummeting to its doom.

Let us substitute this value of v into the earlier formula for centrifugal force to get:

$$F = \frac{m}{r}\left(\frac{L}{mr}\right)^2 = \frac{L^2}{mr^3}$$

This shows that the value of the centrifugal force varies as the inverse **cube** of the distance, r. But we know that gravity only varies as the inverse **square** of the distance. So as the satellite moves closer to the Earth, gravity increases – but the centrifugal force increases quicker. Hence, at a certain distance, the value of the centrifugal force will increase so that it balances gravity again. The International Space Station will be saved!

This analysis has shown that orbits are stable: a slight variation in velocity will not result in a satellite plummeting downwards.

However, as explained earlier, the strength of gravity is strongly dependent on the number of spatial dimensions. So this result suggesting the stability of orbits is only applicable in three dimensions. As explained earlier, if there are more than three dimensions then gravity becomes very much stronger as objects move closer together. This effect would overwhelm the increase in centrifugal force, so nothing could prevent gravity from pulling the satellite to its doom.

In other words, orbits are only stable in three dimensions.

Anthropic theories

The stability of orbits might just be regarded as a fortuitous consequence of three-dimensional space. However, the stability of orbits also has a surprising interpretation: it appears to mean that life can only develop in three-dimensional space. As John Barrow said in *Dimensionality*, "If hundreds of millions of years in stable orbit around the Sun are necessary for planetary life to develop then such life might only develop in a three-dimensional world."

This restriction is also the case for fewer than three dimensions. Max Tegmark, a physicist at MIT, has suggested that intelligent life could not exist in only two dimensions of space because "two nerves cannot cross."[10] This theme was continued in John Barrow's book *The Constants of the Universe* in which he considered the difficulties of life in two dimensions. For example, a two-dimensional creature would be split into two pieces by its digestive system:

---

[10] Max Tegmark, *On the dimensionality of spacetime*, http://arxiv.org/pdf/gr-qc/9702052v2.pdf

If you are not convinced that the evolution of intelligent life requires the existence of planets in stable orbits, perhaps you might be convinced by the fact that the three-dimensional restriction also applies to the orbits of electrons around the atomic nucleus. Atoms simply could not exist if there were more or less than three dimensions of space. How could life develop in a universe with no atoms? With no chemistry? As John Barrow says: "If we assume the structure of the laws of physics to be independent of the dimension, stable atoms, chemistry, and life can only exist in fewer than four spatial dimensions."

Because of life's apparent requirement for three-dimensional space, in 1955, the English cosmologist Gerald Whitrow was the first to propose a radical idea.[11] Whitrow

---

[11] Gerald Whitrow, *Why Physical Space Has Three Dimensions*, British

considered the stability of orbits and realised that the evolution of life would have been impossible if the universe had not permitted stable orbits. For this reason, Whitrow proposed that the universe **had** to have three dimensions, for the simple reason that life could not have evolved in any other type of universe, and so we simply would not exist in order to observe the universe. Whitrow's idea has come to be known as the *anthropic principle.*

It is the case that several other crucial fundamental constants – most notably the *cosmological constant* – appear to be set to life-friendly values. According to the basic form of the anthropic principle, we should not be at all surprised that that is the case. After all, it could be no other way.

In its most basic sense, the anthropic principle merely states the obvious: the universe clearly has to be amenable to life (with the fundamental constants set to life-friendly values) or else we simply would not exist, and so would not be observing the universe. It is virtually impossible to disagree with this form of the anthropic principle, which should be regarded as a tautology.

However, there is a complicating factor. As it appears to be the case that the fundamental constants could conceivably possess values which are not life-friendly, then the emergence of a life-friendly universe appears to be an extremely fortuitous coincidence. And physicists instinctively don't like coincidences. Coincidences appear to indicate some deeper – or external – unknown mechanism.

In an attempt to explain-away the coincidences, some physicists have proposed a particularly controversial version of the anthropic principle. It has been suggested that there

Journal for the Philosophy of Science, vol. 6, No. 21, 1955.

58

are many separate universes – forming a *multiverse* – with the fundamental constants being set to different values in each individual universe. Hence, we should not be surprised to observe life-friendly conditions in our universe.

This version of the anthropic principle eliminates the coincidences: we just happen to inhabit one of many different universes, a universe which must have life-friendly conditions. But there is clearly a price to be paid in terms of the proposed (and unobserved) multiverse of many (possibly an infinity) different unobserved universes. For this reason, you might imagine that this strong form of the anthropic principle would be generally unpopular, generating little interest.

However, for many physicists, and in many particular areas of fundamental research, quite the opposite has happened. Not only has the strong anthropic principle been welcomed, but it is seen as being capable of providing an explanation for many intractable problems: "Why are there three dimensions of space? Because there are an infinity of universes with different numbers of dimensions of space, and humans could have only evolved in one of those universes which had three dimensions."

On that basis, the anthropic principle solves everything! We might as well retire from physics and spend the rest of our lives down the beach. There are no more problems to be solved.

However, I hope you realise that this forms a very unsatisfactory form of "explanation". Indeed, it does not really resemble an explanation at all, not in the true sense of conventional analytical physics. In his book *The Trouble with Physics*, Lee Smolin regrets the adoption of the anthropic principle by some physicists: "because it has been understood for some time that it is a very poor basis for doing science." Joseph Conlon includes a memorable criticism in his book *Why String Theory?*:

While the anthropic principle is not vacuous, it can be seductive. It offers the dangers of the open cookie jar at Fat Camp – the soft route of easy temptation. It also encourages a solipsistic attitude to science. For example, I could ask why the Cuban missile crisis did not end in mutual assured destruction, with a nuclear conflagration that destroyed the world. If I felt sufficiently brazen, I could respond that the answer is the anthropic principle. If a nuclear war had occurred in 1962, my parents are unlikely to have met a decade later, and I would never have been born in 1981. However, many would feel that the fact that I am right now contemplating the marvel of my own existence is not a satisfactory explanation for why Kennedy and Khrushchev managed to avoid taking their respective nations to war.

For this book, I conducted a survey of existing theories as to why there are three dimensions of space. And, unfortunately, I have to say that the anthropic approach dominates these theories. As well as the previously-discussed stability of orbits, other examples of these anthropic theories include:

- In space with four or more spatial dimensions, all stars either collapse into black holes, or disperse. And no stars means no people.[12]

- Only a three-dimensional universe would expand at a rate which would allow the formation of large-scale structures such as galaxies.[13]

- Waves only travel without distortion or reverberation in three-dimensional space. As John Barrow says in *Dimensionality*: "Three dimensional worlds appear to possess a unique combination of properties which enable information-processing and signal transmission to occur via electromagnetic wave phenomena ... This situation has led many to suppose that life could only exist in an odd-dimensional world because living organisms require high-fidelity information transmission at a neurological or mechanical level." In other words, no information processing means no people.

I get the impression that anthropic theories such as these are invoked when a problem seems so difficult – so impossibly insurmountable – that it is hard to conceive of any conventional analytical solution. Anthropic theories could then be thought of as "theories of last resort".

---

[12] John Bechhoefer and Giles Chabrier, *On the fate of stars in high spatial dimensions*, American Journal of Physics 61, 460 (1993).

[13] M. Gasperini, *The cosmological constant and the dimensionality of space-time*, Physics Letters B, Volume 224, June 1989, pp. 49-52.

However, I would suggest it is best never to resort to such theories.

Instead, if we really want to solve the problem of why there are three dimensions of space, we are going to need the highest-quality science. Let us start with our best theory of the geometry of space ...

4

GENERAL RELATIVITY

In November 1907, Albert Einstein was sitting in a chair in the patent office in Bern where he worked as an assistant examiner. According to Einstein: "All of a sudden, a thought occurred to me. If a person falls freely, he will not feel his own weight." This was the thought which started Einstein on the eight-year road to formulating his great theory of general relativity. Later, he was to call this the happiest thought of his life.

In this book, we are considering clues as to why there are three dimensions of space. As general relativity remains our best theory of space and time – a hundred years after it was discovered – in this chapter we will consider general relativity in some depth.

The idea of a free-falling person not feeling their own weight may not seem to be a particularly surprising or inspiring thought. Indeed, it is a concept which seems so familiar to us that we would take it for granted, and think no more of it. However, as Nobel Prize-winning chemist Albert Szent-Györgyi said: "Scientific discovery consists of seeing what everyone else has seen, but thinking what no one else has thought."

Perhaps we can understand Einstein's insight by considering the modern-day equivalent of his falling man thought experiment, which would be a *reduced-gravity aircraft* flight. A reduced gravity flight gives its occupants (sometimes prospective astronauts, but more often simple thrillseekers willing to pay the $5000 price) a brief experience of genuine weightlessness (approximate duration of 25 seconds). The flight involves a steep climb and then a rapid descent.

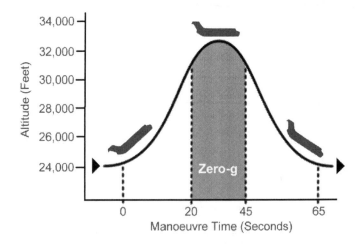

The force of gravity at the bottom of the hump is roughly twice as large as normally experienced. The plane then starts to climb and the sequence starts over again. There are typically 40-60 manoeuvres in a flight. This results in approximately two thirds of the passengers feeling nauseous. Hence, these aircraft have been nicknamed the "vomit comets".

The following NASA photograph shows Stephen Hawking's zero-g flight in 2007.

What the vomit comet (and Einstein's thought experiment) reveals is that it is possible to eliminate the force of gravity purely by placing objects in an accelerated environment. This appears to reveal a fundamental equivalence between gravity and acceleration. This so-called *equivalence principle* means that a passenger in a spaceship in a weightless environment in deep space will feel the force of gravity if the spaceship is accelerated (the passenger will be pushed into their chair).

NASA are considering building a spaceship which generates its own artificial gravity by continuously rotating, rotation being a form of acceleration. However, the equivalence principle tells us that this is no "artificial" gravity – no, **gravity and acceleration are precisely the same thing**. There is nothing "artificial" about this artificial gravity – it genuinely is the force of gravity.

Einstein stated this equivalence principle in 1907: "We assume the complete physical equivalence of a gravitational field and a corresponding acceleration of the reference system."

The curvature of space

The equivalence principle means that if we consider a spaceship in deep space in a weightless environment, and if we then consider that spaceship being accelerated, then the effect on the objects inside that spaceship will be precisely equivalent to the force of gravity. So let us imagine that there is a small window in the side of the spaceship through which light from a nearby star is entering the spaceship. According to the point-of-view of a passenger in the spaceship, the beam of light entering through the window appears to curve to the floor as the spaceship accelerates (as the light crosses the spaceship, it moves closer to the floor as the spaceship accelerates upward).

The following image shows this curved beam of light inside the spaceship:

We know from the equivalence principle that this accelerated spaceship is effectively experiencing the force of gravity. So the light appears to be dropping to the floor under the influence of gravity. This is quite a revelation: the path of light is affected by the force of gravity. Indeed, it was the observed deflection of starlight around the Sun during a solar eclipse which provided the first confirmation of the accuracy of general relativity.

But we know that light always tends to travel in a straight line through space (in fact, as we shall shortly see, light is actually travelling the shortest distance between two points – which usually happens to be a straight line). So if light truly is continuing to travel in a straight line through space under the influence of gravity, then this implies that **space itself must be curved**.

This is the main message of general relativity: **the effect of gravity is caused by the curvature of space** (actually, the curvature of spacetime). We will see later in this chapter precisely how this curvature of space causes the forces we associate with gravity.

In the later years of his life, Einstein explained his thought processes to his youngest son Eduard. Describing his insight that gravity was the curving of spacetime, he said: "When a blind beetle crawls over the surface of a curved branch, it doesn't notice that the track it has covered is indeed curved. I was lucky enough to notice what the beetle didn't notice."

Tensors

Once Einstein realised that gravity was caused by the curvature of space, he was faced with the task of expressing that curvature mathematically (so it could be included in the eventual final equation). Einstein's mathematical skills were good, but he was not a professional mathematician, and he realised he needed help. So, in 1912, Einstein called on his old friend Marcel Grossman.

Grossman had taken the class notes when Einstein had missed the mathematics lectures at the Zurich Polytechnic. According to Walter Isaacson in his biography of Einstein, Einstein had scored 4.25 out of 6 in his geometry class at the polytechnic, while Grossman had scored a perfect 6.

Grossman was very excited when Einstein asked for his help to describe the curvature of space in a mathematical form. Grossman consulted the literature, and discovered that Einstein required *non-Euclidean geometry* to describe the curvature. Euclidean geometry is the conventional geometry taught in schools based on lines, squares, circles, etc. drawn on a flat plane. However, non-Euclidean geometry extends this geometry to consider shapes drawn on curved surfaces. The resultant geometry can appear very counter-intuitive. For example, the interior angles of a triangle can sum to more or less than 180° when drawn on a curved surface, and parallel lines can eventually meet.

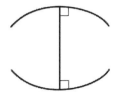

In the previous diagram, we might imagine the sphere as representing the Earth. We can then imagine two explorers at different positions on the Equator, and they both decide to walk in a straight line in a precisely northerly direction (along the lines shown in the diagram). The paths they take will gradually become closer together until they eventually meet at the North Pole. We might interpret this apparent attraction between the explorers as being due to the force of gravity, but really they are just following straight lines along curved space. It is the curvature of space which introduces gravity.

As we move off the two-dimensional plane and consider the possibility of non-Euclidean geometry in higher dimensions, the important factor for describing those spaces mathematically is how we measure the distance between two points. As we shall see, this is because – in curved space – the shortest distance between two points can be different from the distance in flat space.

As was described in Chapter Two, the mathematical formula for the shortest distance between two points is called the *metric*. Firstly, let us see recap how this metric is defined on a two-dimensional flat surface.

In the following diagram, imagine we want to travel from point A to point B via the shortest possible route (the shortest possible route between two points in geometry is called a *geodesic*). The shortest distance between the two points is shown by the dotted line. You will see that travelling from point A to point B necessitates travelling a distance x horizontally, and a distance y vertically:

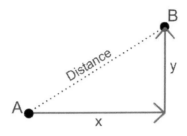

This forms a right angled triangle, so from Pythagoras's theorem we could calculate this distance as the square root of x^2 plus y^2. So this gives us our metric on a flat two-dimensional plane:

$$d = \sqrt{x^2 + y^2}$$

However, this metric is not necessarily the same for curved surfaces. In order to see why that can be the case, consider an aircraft flying from Seattle to Zurich (see the following diagram). Clearly, the airline will want the aircraft to take the shortest possible route (to save fuel). It so happens that Seattle and Zurich both lie at the same latitude (47° North), which implies that Zurich is precisely east of Seattle. We might therefore expect the aircraft to fly in a precisely easterly direction (which we might consider to be the horizontal x direction). On flat space, this would be correct: in the following diagram we would expect the aircraft to follow one of the many circular lines of latitude which are drawn in light grey around the globe:

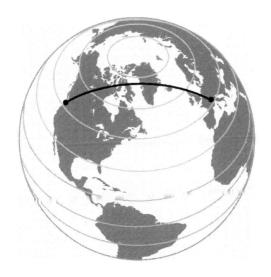

However, because the surface of the Earth is curved, you will see on the diagram that the route the aircraft takes is not the easterly route along one of the lines of latitude. Instead, the aircraft takes a more northerly route, as far north as Greenland (the route is shown by the curved black line on the diagram).[14]

So, whereas on a flat surface, the shortest distance between Seattle and Zurich would simply have been the distance in the easterly (x) direction, the situation is more

---

[14] This flight pattern – which goes so far North – is called a *great circle*. Unfortunately, the *Titanic* hit an iceberg because it was following a northerly great circle route (the fastest route to New York) which took it into an ice field.

complicated for curved surfaces. On a curved surface, the shortest route might be very different.

Grossman realised that Einstein would need to use *tensors* to describe curvature in his eventual equation for general relativity. A tensor is similar to a *vector*, and vectors were introduced in my previous book. A vector is a column matrix – a matrix (a rectangular array of numbers) which has just one column:

$$\begin{bmatrix} x \\ y \\ z \end{bmatrix}$$

In my previous book, we saw how vectors can be used for representing points in space. For example, the vector shown above could be used to represent the point with the coordinates (x, y, z).

A tensor is the more general case of the vector. For our purposes, we can consider a tensor as being a two-dimensional array of numbers. For example, a flat surface is represented by the following *metric tensor*:

$$\begin{bmatrix} 1 & 0 \\ 0 & 1 \end{bmatrix}$$

A metric tensor allows us to compute the distance between two points in any space.

So why is this the metric tensor which describes a flat surface? Well, let us imagine you are an aircraft pilot, and you want to know how far you will actually travel (and therefore how much fuel you will use) if you travel a certain distance in the easterly (x) direction, and a certain distance in the northerly (y) direction. On a flat surface, the formula which calculates the distance you will actually travel – the metric – is given by:

The metric tensor

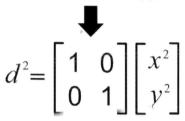

$$d^2 = \begin{bmatrix} 1 & 0 \\ 0 & 1 \end{bmatrix} \begin{bmatrix} x^2 \\ y^2 \end{bmatrix}$$

This formula is in the form of a matrix multiplication (a tensor multiplying a column matrix), so let us perform the matrix multiplication and see what we get (the technique of matrix multiplication was described in my previous book).

When we multiply the matrices, we get:

$$d^2 = x^2 + y^2$$

which means:

$$d = \sqrt{x^2 + y^2}$$

which, as we saw previously is the metric for a flat two-dimensional plane (which we calculated previously from Pythagoras's theorem).

So we have, indeed, discovered the metric tensor for a flat surface, and it is:

$$\begin{bmatrix} 1 & 0 \\ 0 & 1 \end{bmatrix}$$

But space has three dimensions. What if we want to calculate the distance travelled (and the fuel used) by the aircraft as it not only moves along the surface, but also flies up into the air? In that case, we have to consider a third (z) dimension, and our metric tensor becomes a 3×3 matrix:

$$d^2 = \begin{bmatrix} 1 & 0 & 0 \\ 0 & 1 & 0 \\ 0 & 0 & 1 \end{bmatrix} \begin{bmatrix} x^2 \\ y^2 \\ z^2 \end{bmatrix}$$

You will see the metric tensor for three-dimensional flat space has a similar form to the two-dimensional case (a *diagonal matrix*: all the elements which are not on the main diagonal are set to zero).

But Einstein knew that special relativity implied that the universe is not built on just the three dimensions of space – it is built on the four dimensions of spacetime. So Einstein knew that the metric tensor describing the curvature of four-dimensional spacetime would have to be a 4×4 matrix.

But, before we consider the necessary matrix, we have to ask the question: how do we actually measure distance in spacetime? Well, the distance travelled by an observer in spacetime can be measured simply as the elapsed time measured on a clock carried by that observer (this is called *proper time*).

This has an interesting consequence: even for an observer who is stationary in space, the clock will continue to show the passing of time, meaning that the observer is actually travelling in spacetime. As explained in my third book, even a stationary observer is actually moving at the speed of light in spacetime. This means that, if the clock of a stationary observer shows that a time t has passed, the observer will actually have travelled a distance ct in spacetime (where c is the speed of light).

There is an additional factor we have to consider. According to special relativity, an observer who travels at high speed (for example, close to the speed of light) will experience less time passing than an observer who remains stationary. This is the principle of *time dilation*. For example, consider an astronaut who flies off from Earth reaching nearly the speed of light. When that astronaut returns to Earth, he will have aged less than an observer who has remained on Earth. Special relativity has therefore revealed that the greater the distance you travel through space, the less distance you will travel through time – and vice versa. As far as the metric tensor is concerned, this means the element of the tensor which multiplies the time component must be the opposite sign (positive or negative) to the sign of the element which multiplies the space component.

To see this, the following diagram shows the metric tensor needed to describe flat (uncurved) spacetime. You will see that the single element in the tensor which multiplies the time component is positive (+1), whereas the three elements which multiply the space components are negative (-1):[15]

The metric tensor

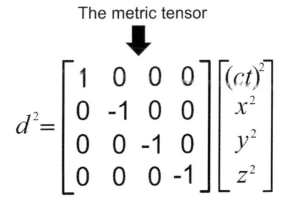

$$d^2 = \begin{bmatrix} 1 & 0 & 0 & 0 \\ 0 & -1 & 0 & 0 \\ 0 & 0 & -1 & 0 \\ 0 & 0 & 0 & -1 \end{bmatrix} \begin{bmatrix} (ct)^2 \\ x^2 \\ y^2 \\ z^2 \end{bmatrix}$$

When we perform the matrix multiplication, we get:

$$d = \sqrt{(ct)^2 - x^2 - y^2 - z^2}$$

---

[15] This sign convention (+−−−) is sometimes called the *West Coast metric*. This is to distinguish it from the (−+++) sign convention which is sometimes called the *East Coast metric*. Some would say either convention is equally valid, but the West Coast metric clearly makes more sense to me as it applies to objects with a *timelike* world line, i.e., objects which cannot move faster than light (which is the case for all objects). In contrast, the East Coast metric applies to events with *spacelike* separation.

which is the correct formula for distance in a flat (uncurved) spacetime. For a curved spacetime, the metric tensor would be different and so the formula for spacetime distance would also be different.[16]

What is more, **all** observers – no matter how they are moving – will agree on the same measured spacetime distance, d. Einstein realised this property was essential: relativity tells us that all frames of reference should be considered equally valid. No observer can claim that his frame of reference (and any single set of coordinate axes) are preferred over any other observer's frame of reference (in Chapter Two it was explained how there can be no absolute time and space in the universe, no single reference for time and space). By using this tensor formulation, the equation would apply to all observers – no matter how they were moving. As we have just seen, it is possible for tensors to produce invariant values on which all observers can agree. Therefore, no single observer – and no single set of coordinate axes – is special or preferred in any way.

Einstein called this invariant property of tensors *general covariance*. Much of the eight years Einstein spent developing general relativity was spent ensuring his equation had this property of general covariance.

-----

[16] Actually, this formula – which measures spacetime distance as the time measured by a travelling clock – means that objects move so as to **maximise** their distance through spacetime, not **minimise** their distance. This is because objects moving along a geodesic will show the maximum time on their clock (whereas as astronaut who deviates from the geodesic – by travelling to Jupiter and back, for example – will show less time passing, less spacetime distance, on his clock when he returns). But the end result is the same: objects tend to travel along geodesics.

Tidal forces

As we saw earlier in our discussion of the equivalence principle and the "vomit comet", it is possible to virtually eliminate the effect of gravity if an object is in free-fall. However, it is actually quite difficult (effectively impossible) to completely eliminate the effect of gravity. That is because space possesses curvature, and that curvature means objects never fall in precisely the same direction – even in a vomit comet. In other words, the geodesic lines of objects never point in exactly the same direction. It is very difficult to choose a region in which space is completely flat (uncurved). But by choosing a very small region – for example, the interior of a vomit comet – space will be approximately flat. To see why, consider a very small region on the surface of a sphere. A small region will be approximately flat (for example, we feel the Earth is flat when we inhabit just a small region of its surface), but there will still be a small degree of curvature present in that small region which can never be completely eliminated.

This ever-present curvature of space can never be completely eliminated in an accelerated frame. The gravitational effects of this curvature are, therefore, the true effects of gravity, and it is these effects we want to consider. These effects are called *tidal forces*.

These tidal forces emerge when objects fall along lines (geodesics) which point in **different** directions – not the same direction. Hence the saying "gravity is caused by the **curvature** of spacetime".

In the following explanation of tidal forces, we will be considering the movement of a number of footballs (soccer balls) as they each follow their geodesic lines through space.

The resultant deformation in the arrangement of the footballs will reveal the action of gravity on structures.

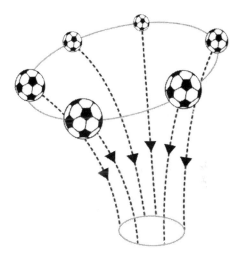

To see the effect of these tidal forces (due to curvature), we need to consider a larger region of space. So consider an observer inside an extremely large elevator, floating freely in space a few hundred miles above the surface of the Earth. Not only will the observer be floating in a weightless situation, but any objects with him in the elevator will also be floating. Imagine there are two footballs in the elevator with the observer. These are floating by the side of the observer, with the balls separated by a certain distance. As was explained earlier, by considering the motion of these footballs in free-fall, we can learn about the behaviour of gravity.

As the elevator continues on its free-fall journey toward the Earth's surface, both the observer and the footballs will be attracted toward the precise centre of the Earth. The footballs follow their geodesics towards the centre of the Earth. This will inevitably result in the paths of the balls slowly converging (see the following diagram – the geodesics of the footballs are denoted by dashed lines). From the point of view of the observer, he will see the balls floating on either side of him, and then he will see those balls slowly drifting together as they follow the converging geodesics:

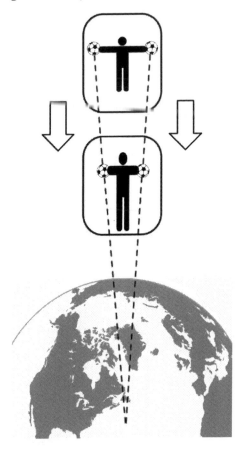

As well as the balls slowly drifting together, there is also a stretching of the observer in the vertical direction (also shown on the diagram). This is due to the difference in the strength of gravity in the vertical direction. This stretching effect is called *spaghettification* and is the effect which will kill you if you are unlucky enough to fall into a black hole. Remember: in free-fall (falling into a black hole, for example) you do not feel the effect of gravity – you only feel the tidal forces, and it is those tidal forces which will rip you to shreds.

In order to discover the effects of these tidal forces on three-dimensional objects, let us replace our observer in the elevator with a very large rubber weather balloon. The footballs are now glued around the edge of the balloon, so that the movement of the footballs (in free-fall along their geodesics) controls the shape of the balloon.

As in the previous example with the observer, the effect of the tidal forces is to squeeze the balloon horizontally, while stretching it vertically. The balloon therefore forms an elliptical shape:

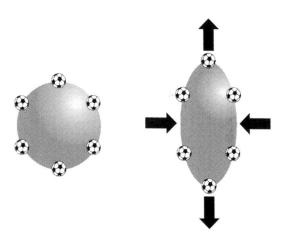

Crucially, even though the balloon forms this elliptical shape, **the volume of the balloon does not change**. Even though the balloon is squeezed around the middle, it is stretched at its ends, and the end result is a volume which is unchanged. Another crucial point is that there is nothing – no mass – inside the balloon (apart from air). So in this situation in which the balloon is empty, the volume of the balloon does not change due to the tidal forces.

Now let us put some mass in the balloon. In fact, we will put a lot of mass in the balloon. Imagine a balloon considerably larger than the Earth, and let us put the Earth inside the balloon. Again, some footballs are glued around the edge of the balloon so that – as the footballs fall along their geodesics – we can observe the effect on the shape of the balloon.

As you can see from the previous diagram, the footballs all fall along their geodesics (straight lines directed to the centre of the Earth). So, in this case, there **is** a volume reduction in the size of the weather balloon.

To recap: if there is no mass inside the weather balloon, the effect of the gravitational tidal forces does not change the volume of the balloon. But if there is mass inside the balloon, the effect of the tidal forces is to reduce the volume of the balloon.

Einstein realised he would have to capture this volume-reducing effect of gravity (due to the presence of mass) in his equation for general relativity. The tensor which can be used to describe volume reduction of an element of space is called the *Ricci tensor* (pronounced "Reechy") which we will denote by $R_{\mu\nu}$.

The rate of volume reduction (described by the Ricci tensor) is then proportional to the mass inside that volume. The distribution of mass (and therefore also the distribution of energy via the conversion equation $E=mc^2$) is described by the *energy-momentum tensor*, $T_{\mu\nu}$. This gives us the following equation:

$$R_{\mu\nu} = k\, T_{\mu\nu}$$

where k is some proportionality factor. It turns out that k must be equal to $8\pi G$ (in order to agree with Newtonian gravity). So the equation then becomes:

$$R_{\mu\nu} = 8\pi G T_{\mu\nu}$$

This equation was Einstein's initial proposal for the general relativity equation. Essentially, what it is telling us is something quite simple: it is telling us that the curvature of space (spacetime) is proportional to the mass (and energy) contained within that volume of space.

However, Einstein realised there was a problem with this initial proposal for a general relativity equation. The problem was due to the law of conservation of energy. Let us first consider the right hand side of the equation, which deals with the distribution of mass and energy. The law of conservation of energy states that the amount of energy in a closed volume of space must remain constant over time. The right hand side of the equation satisfies this law because if we consider the derivative of the energy-momentum tensor (the derivative is the rate of change of a value over time) then we find that it is equal to zero: the amount of energy in the volume does not change over time, i.e., energy is conserved. We can express this as:

$$\nabla T_{\mu\nu} = 0$$

where the ∇ symbol represents the derivative.[17]

However, the derivative of the Ricci tensor is not zero, so Einstein's equation could not be correct (the derivative of the right hand side of the equation was zero, but the derivative of the left hand side of the equation was not equal to zero). Instead, the derivative of the Ricci tensor, is:

$$\nabla R_{\mu\nu} = \tfrac{1}{2} R \nabla g_{\mu\nu}$$

where R on the right hand side is called the *Ricci scalar* (it is derived from the Ricci tensor, and it is just a number which is not zero if the surface is not flat), and $g_{\mu\nu}$ is the metric tensor which we considered in the previous section.

Einstein realised he could take the right hand side of the equation over to the left hand side, changing its sign in the process, and taking the derivative outside the bracket (as it was a factor which was common to both terms):

$$\nabla(R_{\mu\nu} - \tfrac{1}{2} R g_{\mu\nu}) = 0$$

We see we are now taking the derivative of the expression inside the bracket, and the result of that derivative is zero. We saw earlier that the derivative of the

---

[17] More precisely, this represents the *divergence*, which is a measure of how quickly a property "spreads out" from a point in space.

energy-momentum tensor is also equal to zero. So Einstein realised that these two terms (on which the derivative was being applied) were equivalent. This meant:

$$R_{\mu v} - \tfrac{1}{2} R g_{\mu v} = T_{\mu v}$$

We also need to include the previous $8\pi G$ proportionality factor:

$$R_{\mu v} - \tfrac{1}{2} R g_{\mu v} = 8\pi G T_{\mu v}$$

And that's it! That is Einstein's equation for general relativity, the derivation of which is generally considered to be the greatest scientific achievement in the history of humanity. It is called simply the *Einstein equation* (yes, that's correct, $E=mc^2$ is **not** the "Einstein equation").

So, although the equation looks rather daunting and mathematical, we can see from our earlier analysis that it is actually telling us something which is really quite simple. In a simplified form, it is telling us that an amount of mass (on the right hand side of the equation) will result in curvature of spacetime (the left hand side of the equation) around that mass. And that, basically, is the essence of gravity.

In his biography of Einstein, Water Isaacson put Einstein's achievement into perspective:

Einstein had shown that the fabric of spacetime became not merely a container for objects and events. Instead, it had its own dynamics that were determined by, and in turn helped to determine, the motion of the objects within it. The curving and rippling fabric of spacetime explained gravity, its equivalence to acceleration, and, Einstein asserted, the general relativity of all forms of motion. In the opinion of Paul

Dirac, the Nobel laureate pioneer of quantum mechanics, it was "probably the greatest scientific discovery ever made". Another of the great giants of twentieth century physics, Max Born, called it "the greatest feat of human thinking about nature, the most amazing combination of philosophical penetration, physical intuition and mathematical skill."

On the largest of scales (which is where general relativity's modification of Newton's law becomes measureable), general relativity correctly predicted *gravitational lensing* (light bent around stars and galaxies), black holes, and the expanding universe.

The final confirmation of general relativity made global headlines recently. General relativity states that gravitational effects are not transmitted instantaneously (after all, nothing travels faster than light). Instead, gravity is transmitted across the universe in the form of *gravitational waves*. As an example, if the Sun disappeared instantaneously, the Earth would continue to orbit the position of the Sun for seven minutes until the bad news reached us in the form of gravitational waves.

Gravitational waves are predicted to carry energy, just like any other form of wave (you may have felt the force of waves in a rough sea). Unfortunately, gravitational waves are also predicted to be extremely weak and very difficult to detect. As an example of the weakness of gravitational waves, the largest source of gravitational waves in the Solar System is due to the orbit of Jupiter around the Sun, but the amount of energy emitted by Jupiter in the form of gravitational waves is only about that of a 40-watt light bulb.

However, despite their weakness, several teams of researchers have spent decades searching for gravitational waves using highly inventive experiments and sensitive detectors. The most sensitive is the Advanced LIGO project (Laser Interferometer Gravitational Wave Observatory) which consists of two observatories based in Livingston in Louisiana and Hanford in Washington State. The experimental equipment at the two sites consists of two four-kilometre long tubes oriented at ninety degrees. Laser beams are shone down both tubes, reflected at the ends 280 times (giving a combined distance of over 1,000 kilometres), before being combined back at the base station. As a gravitational wave passes, it is predicted to slightly shorten the length of one of the tubes while slightly lengthening the other (because the tubes are at ninety degrees). The difference in the path length of the laser beams would then result in an interference pattern when the two laser beams are recombined at the base station.

The following picture shows the LIGO observatory in Hanford. You can see the two tubes oriented at ninety degrees:

The search bore fruit at 9:50 a.m. on the 14th September 2015 when the two Advanced LIGO observatories independently detected a brief burst of gravitational waves from the collision and merger of two black holes. As the gravitational wave passed through the Earth, it changed the length of the four-kilometre LIGO tube by just ten thousandth of the width of a proton, equivalent to detecting a change in the distance to the nearest star by the width of a hair.

At its peak, the energy of the gravitational waves produced by this cataclysmic event was 50 times greater than the combined power of all the light radiated by all the stars in the observable universe. Effectively, the mass of three Suns was converted into raw energy.

If the energies involved were so huge, then why was the signal detected on Earth so weak? To put it simply, the black hole merger happened in a galaxy far, far away and a long time ago. The merger actually happened 1.3 billion years ago (in a galaxy 1.3 billion light years away, logically). At that time, the only life on Earth was composed of simple single cells floating in the oceans. The gravitational wave started its journey across space to be intercepted – 1.3 billion years later – by multi-cellular human beings which had fortunately evolved over that period.

The following graphs show the "chirp" signal received from the black hole merger which lasted just 0.2 seconds – approximately the same time as the blink of an eye. The first column shows the signal received in Livingston, the second column shows the signal received in Hanford just seven milliseconds later (the time it takes for light to travel across the USA).

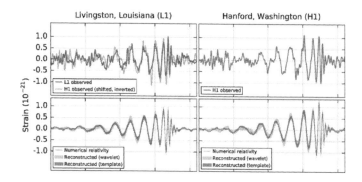

The lower graphs show how the signal was predicted to appear (including the so-called black hole "ringdown" after the merger), showing how well general relativity correctly predicted the nature of the event.

The following diagram shows how the merger of the two black holes related to the detected "chirp". Remember, this whole process took just 200 milliseconds to complete. The black holes were orbiting each other at a distance at which they were only 350 km apart. Over the 200 millisecond period, the orbital velocity of the black holes increased from 30% of the speed of light to 60% of the speed of light.

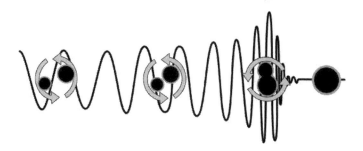

This represented the first direct detection of gravitational waves, and is regarded as one of the most important discoveries in the history of science.

5

INTO THE FIFTH DIMENSION

When you have made the greatest scientific discovery of all time, the question then becomes: "How do I follow that?" That was surely the challenge facing Einstein after he discovered general relativity in 1915. However, his course of action probably appeared quite clear. Just as is the case today, there were a few fairly clear problems in fundamental physics which cried out for solutions. And, just as is the case today, these questions tended to revolve around the problem of unification.

As described in my first book, if we find we have two theories which explain two apparently different behaviours of Nature, we might find it possible to replace both those theories with a single theory which still manages to explain both of those behaviours. When this happens, the process is called *unification*. The resultant unified theory will be simpler than either of the two previous theories.

As was described in Chapter Two of this book, one way of combining two apparently different behaviours is to introduce extra dimensions (remember the "flying octopus"?). In this chapter we will be examining the first attempt at unification using extra dimensions.

Today, one of the greatest quests in physics is to find a unification between general relativity and quantum mechanics. These are the two dominant theories of the physical world, and the unification of these theories stands out quite clearly as a problem in urgent need of a solution.

In Einstein's time, there was also a very clear need for a unification, a unification between the theory of electromagnetism and Einstein's own theory of general relativity. In order to understand why this was such an important quest, we need to imagine travelling back in time to the early decades of the 20th century, and putting ourselves into the mindset of the physicists of that period.

In the 19th century, the greatest advance in physics since the time of Newton was achieved. This was the development of the theory of *classical* (i.e., non-quantum) *electromagnetism*. In 1820, French physicist Andre-Marie Ampère had shown that wires carrying electric current generated a magnetic field, and those wires were either attracted or repelled by the magnetic force between them, depending on the direction of the electric current. In 1831, Michael Faraday discovered the inverse of this result by demonstrating that a magnet moving through a coil of wire produces an electric current in that wire. So there was clearly a symmetry between electricity and magnetism. It was then left to the great Scottish physicist James Clerk Maxwell who, in 1862, formalised this unification between electricity and magnetism in mathematical form. Maxwell published four equations, now simply known as *Maxwell's equations*.

Of particular interest to Einstein had been the form of Maxwell's equations in a vacuum (empty space containing no electric charges or current). Maxwell had shown that his equations made the symmetry between electricity and magnetism clear:

1) A changing magnetic field creates an electric field.

2) A changing electric field creates a magnetic field.

Hence, the magnetic field can generate an electric field, and the electric field can generate a magnetic field. The result is a self-sustaining wave of alternating electric and magnetic fields called an *electromagnetic wave*. This wave was capable of traversing the vacuum, and the equations implied that the wave would travel at precisely the speed of light. From this, Maxwell deduced that light was a form of electromagnetic wave. What is more, the equations seemed to indicate that the speed of light would be independent of the speed of the observer. Einstein took this result as the basis for his theory of special relativity, with his landmark 1905 paper being called "On the **electrodynamics** of moving bodies". Hence, with special relativity Einstein was unifying Maxwell's new electromagnetic theory with the existing theories of mechanics at speeds close to the speed of light. This was a clear first sign of the need to unify the new results from electromagnetism with existing theory, a situation which only became more pronounced with the discovery of general relativity ten years later.

What is more, we have to realise that in 1915 the secrets of the atom (specifically, the forces contained within the nucleus of the atom) had not yet been revealed. So the electromagnetic and gravitational forces were the only two forces known. The clear challenge was to unify electromagnetism with general relativity.

Einstein was particularly drawn to the idea of finding a geometrical explanation of the electromagnetic force. If gravity was simply the result of curvature of spacetime, might it be possible that the electromagnetic force also had a similar explanation? Einstein might even have suspected that the solution would be rather simple and intuitive (remember his simple idea of a free-falling person not experiencing the force of gravity?). Einstein explained his goal in his Nobel Prize acceptance speech:

> *The mind striving after unification cannot be satisfied that two fields should exist which, by their nature, are quite independent. We seek a mathematically unified field theory in which the gravitational field and the electromagnetic field are interpreted only as different components or manifestations of the same uniform field.*

Unfortunately, Einstein's efforts were to end in failure. However, the effort to produce a unified theory introduced some new ideas which resonate to this day. One of those ideas has emerged as the dominant theme in fundamental theoretical physics over the last thirty years: the introduction of extra spatial dimensions. We will now consider the revolutionary theory which has formed the template for these higher-dimensional theories.

In other words, we will now consider the theory which formed the template for string theory.

Kaluza-Klein theory

In 1919, Einstein received a letter from an obscure German mathematician named Theodor Kaluza. In today's world, the letter would probably have been dismissed as the ramblings of a "crackpot" – the central idea was so bizarre and apparently nonsensical. However, Einstein had a more open mind, and he gave serious consideration to the contents of the letter.

According to Brian Greene in his book *The Elegant Universe*: "Kaluza's suggestion has revolutionized our formulation of physical law. We are still feeling the aftershocks of his astonishingly prescient insight."

In the letter, Kaluza described an approach for unifying the gravitational force and the electromagnetic force. In order to achieve this, Kaluza had considered how gravity would behave if there were five spacetime dimensions instead of four dimensions. Yes, it sounds like a crazy idea – the universe surely only has four spacetime dimensions? However, the additional dimension seemed to introduce behaviour which was identical to the behaviour of electromagnetism. Could this really be the solution to unification? Does the universe really have five spacetime dimensions?

In order to achieve this apparent miracle, Kaluza did nothing particularly remarkable: he simply considered what would happen if another spatial dimension was added to Einstein's equation for general relativity. As we have seen in the last chapter, the metric tensor used in general relativity is a 4×4 matrix. With the addition of an extra space dimension, Kaluza, therefore, had to increase the size of the metric tensor to a 5×5 matrix (which could now be used for calculating distances in five dimensions):

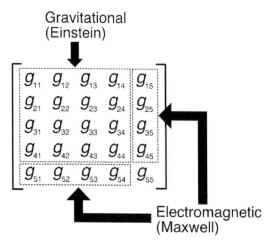

You will see in the previous diagram that the old 4×4 matrix of Einstein's general relativity is retained (in the top left). However, the larger size of the new metric has introduced two new vectors containing four elements (a metric tensor is actually symmetrical, so these two new vectors are actually identical).[18]

Crucially, Kaluza discovered that this new vector represented the laws of electromagnetism. So it appeared that by expanding general relativity to four space dimensions, electromagnetism was automatically incorporated. Hence, the gravitational and electromagnetic forces would be unified if there were four spatial dimensions (five dimensions in total, when the time dimension is also included).

So where does this connection with electromagnetism come from? Well, let us consider what we have done. We have extended Einstein's metric tensor by adding an extra space dimension. We know that Einstein's metric tensor for general relativity described a universe with one dimension of time and three dimensions of space. We have added an additional dimension of space, and produced a vector with four elements. We would therefore expect just one of the elements in this new vector to be a time component (as was the case in the old metric tensor). And this is the case. The new vector has one time component, and three space components:

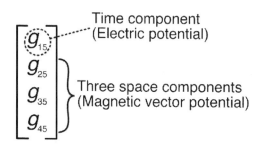

The structure (and behaviour) of this vector is well-known in electromagnetic theory and is called the *electromagnetic vector potential* (alternatively called the *electromagnetic four-potential*). The single time component is a single number and this is called the *electric potential*. The three remaining components form a vector which is called the *magnetic vector potential*.

So clearly this vector combines both electric and magnetic behaviour in a single vector. In other words, it describes a combined **electromagnetism.** So we appear to have obtained a description of electromagnetism purely by extending general relativity to five dimensions.

And that, basically, is the principle behind Theodor Kaluza's unification of electromagnetism and gravity.

Can we get a deeper understanding of this connection between electromagnetism and gravity? Well, the connection appears to rely on the vector which was produced: the electromagnetic vector potential. That vector included the electric potential, and the magnetic vector potential. I hope you can see a common theme here: it is this property called **potential**, which keeps appearing.

It would appear that in order to obtain a deeper understanding of the connection between gravity and electromagnetism we will have to explore this property called "potential". In particular, we have to examine why it appears to suggest the existence of an extra dimension of space.

Potential

As a first step toward unification of electromagnetism and gravity, we might ask what they have in common. Well, both forces are transmitted via fields. It was Michael Faraday in 1849 who first suggested that the magnetic and electric fields were independent entities which spread through space. The movement of an object was then determined by the local magnitude and direction of the field at any point in space. Einstein applied similar reasoning in general relativity. Gravitational effects could not be transmitted instantaneously (as suggested by Newton). Instead, objects responded to the local strength of the gravitational field, and gravitational effects were transmitted through the field at the speed of light (gravitational waves).

So what determines the strength and direction of the field (electromagnetic or gravitational) at any particular point in space? To answer that question we need to consider energy.

If we move an object against the natural direction of a field (for example, pushing two electrically-charged particles together, pushing two north poles of two magnets together, or lifting an object against the force of gravity) then we increase the *potential energy* of the object. If you then release the object (for example, dropping a rock you have lifted off the floor), the force will act to move the object back to its initial position. The potential energy of the object will then decrease. Because of the law of conservation of energy, energy can never be created or destroyed: it can only be converted into a different form of energy. Hence, the kinetic energy of the object will increase accordingly: the motion of the object will accelerate.

We can think of potential energy as being an energy difference between the current position of the object and the position the object would attain if it was released. The greater the difference, the greater the potential energy. For example, the gravitational potential energy of an object is proportional to the height the object is lifted off the ground. So potential energy is clearly an energy associated with the position of an object. And if an object is in a force field (gravitational or electric) then the potential energy of the object is related to the position of the object in that field.

We can therefore consider the field independently and say that a particular point in the field represents a certain amount of potential energy. We have to consider the potential energy which would be possessed by an object with a mass of just one unit (or electric charge of just one unit) at that particular position in the field. We can then say that a particular point in the field has a certain *potential*.

But we want to discover what determines the strength and direction of the force experienced by the object. We will now show that objects will tend to move from points in the field which have high potential to points which have low potential.

To illustrate this, we shall consider a very useful analogy between the flow of electricity and the flow of water under the influence of gravity. Let us consider the potential energy of water in a hydroelectric plant. Water from a reservoir or river flows down a tube to a lower height, until it hits a turbine at the bottom of the tube. The force exerted by the water is sufficient to turn the turbine and generate electricity.

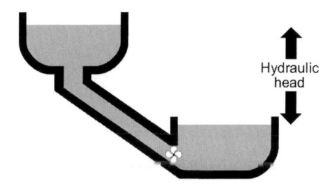

As any civil engineer will know, the force (pressure) exerted by the water on the turbine is proportional to the height difference between the top surface of the water in the lake and the height of the turbine. This is called *hydraulic head*. Hydraulic head is equivalent to the potential energy of the water. And, as we have just discussed, there is a direct connection between potential energy and the potential of the underlying field (in this case, the gravitational field). We can see that water will flow from points with high gravitational potential to points with low potential.

In this respect, there is a perfect analogy with electricity. This is because another name for electrical potential difference is *voltage*. Electricity flows from high potential to low potential, from high voltage to low voltage. So, in our quest to unify electromagnetism with gravity, here we have another similarity between the electromagnetic force and the gravitational force: both forces tend to move objects from high potential to low potential.

In our discussion of the hydroelectric plant, we can see that the only important factor which determines the strength of the force is the potential **difference** (the difference in the height of the water). This seems to indicate that it is the

relative change in potential which is important – not the **absolute** value of the potential.

As an example, it has been mentioned that another name for a difference in electrical potential is **voltage**. So let us consider a typical 1.5 volt battery:

When you buy a battery you are, of course, buying a battery with two terminals: a positive terminal and a negative terminal (shown on the diagram above). The 1.5 volts refers to the potential difference between the two terminals.

So let me ask you a question: what is the voltage of the positive terminal? You might say, "positive 1.5 volts". Are you sure? Well, if you're sure then I will just remove the negative terminal altogether and sell you a battery with only the positive terminal. After all, the 1.5 volts of electricity provided from the positive terminal should be enough to power your device:

I hope you will be very happy with your new battery. I will even sell it to you for half price.

Of course, this is a nonsense. Your new battery with its "1.5 volt terminal" is now incapable of powering anything. A battery with one terminal is useless – it no longer has a voltage. Remember, a voltage is a potential **difference**. When you buy your battery, you are buying a potential difference, and a difference requires two terminals.

If the only important factor is the potential difference then that implies that the absolute value of the potential does not matter. To return to our water analogy, it implies that the absolute height of our hydroelectric power station is irrelevant: the only important factor is the difference in the height of the water, not the absolute value of the height of the water. Hence, we could locate our hydroelectric power station at ground level, or at the top of a mountain, or at any height in between: the potential difference of the power station would be unchanged, and hence the output of the power station would be unchanged.

As that is the case, then imagine you are provided with the value of the output from a particular power station on the mountain. It is clear that there is no clue in the output value as to the absolute height of the station which provided the power. In fact, if the only information available to us is the power output of the station, then this freedom of motion of the station in the vertical direction would be effectively hidden from us. In other words, there is a freedom of motion of the absolute value of the potential: the absolute value of the potential is hidden from us.

Similarly, because the electromagnetic force is also dependent on potential **difference** (voltage) there is also a freedom of the **absolute** electromagnetic potential which is fundamentally hidden from our eyes: it is unobservable.

As was discussed in Chapter Two, it is possible to express the variation of any property value as movement in a dimension. On that basis, it is possible to describe the freedom of variation of absolute electromagnetic potential as movement in a dimension. And, as variation in the absolute value of potential is fundamentally hidden from our eyes, this has to be movement in a **hidden dimension**.

So now we can see the connection between potential and motion in a fourth dimension of space, which was revealed by Kaluza's theory. If you remember back to the discussion earlier in the chapter, Kaluza made the radical suggestion that this motion of absolute electromagnetic potential in a hidden dimension actually represented motion is a real dimension of space! Kaluza showed that it was possible to extend general relativity to five spacetime dimensions, in which case the extra dimension of space behaved like electromagnetic potential. We can now see that this additional fourth dimension of space would be hidden from our eyes.

But what form might this dimension take? How can there be an extra dimension of space which we are fundamentally unable to observe?

The circular dimension

Remember that even though Kaluza's theory was intriguing, there was still absolutely no evidence of a fourth spatial dimension. Theodor Kaluza was a firm believer in the power of theory (he taught himself to swim by reading a manual and jumping into the sea), but in this case theory did not appear to match reality.

However, in 1926 Felix Klein came up with a proposal which not only described the shape of the additional spatial dimension, but also explained why it was hidden from our eyes.

What Klein managed to do was identify the true nature of the connection between the gravitational force and the electromagnetic force indicated in Kaluza's theory. They are both an example of a *gauge theory*. Those of you who read my previous book will already be aware of gauge theory. It is currently believed that all four fundamental forces (gravity, electromagnetic, and the weak and strong nuclear forces) are examples of gauge theory.

Gauge theories arise from symmetry in Nature. A symmetry occurs when you transform an object in some way, but the transformation leaves the original object unchanged. For example, a snowflake has rotational symmetry: if you rotate a snowflake by sixty degrees then the transformed snowflake remains identical to the original snowflake, and that represents a symmetry. The gravitational force is the force which describes the behaviour of space and time, and we find symmetries in both space and time: if we move an experiment to a different location in the universe, we find we get the same result. This is called *space translation invariance* (it is the freedom we have to locate our power station anywhere in the vertical direction). And if we perform an experiment at a different time, we also find we get the same result (this is called *time translation invariance*).[19]

But what symmetry does the electromagnetic force represent? Well, once again the answer was described in my previous book. We will now discover that the symmetry represented by the electromagnetic force is the rotational symmetry of a circle.[20] To understand why this is the case, we need to consider quantum mechanics.

Quantum mechanics tells us that an electrically-charged particle (such as an electron) is described by a *wavefunction*. A wavefunction – as the name implies – has a shape like a wave. And, like any wave, the wavefunction has an

---

[19] These are examples of *global symmetries* because the same symmetry transformation is applied at every point in the universe. Gauge theory arises from *local symmetry* in which different transformations can be applied at each point in the universe.

[20] Technically, this is called *U(1) symmetry*, which means "rotation in one complex dimension". For more information, see the book *Deep Down Things* by Bruce Schumm.

associated *phase*. The phase of a wave determines where the peaks and troughs of the wave lie.

As you can see in the following image, we can imagine a wave being created by a point on a revolving wheel (the revolving wheel is shown on the left in the following image):

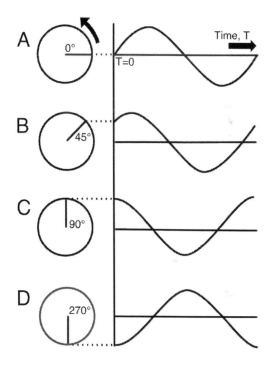

The phase of the wave is then defined as the angle of the wheel when time (on the horizontal axis) equals zero, as shown in the diagram above.

You will see that the last two waves, waves C and D, have phases of 90° and 270° respectively. Hence, there is a 180° difference in their phases. If these two waves were to combine then they would cancel each other out (destructive interference) because the peaks and troughs of one wave would be precisely the opposite of the other wave.

Quantum mechanics tells us that if we have a group of particles – and their wavefunctions – then we can increase or decrease the phase of all the wavefunctions by a constant angle without changing the overall situation. This is because all that matters is the difference of the phases of the wavefunctions, not the absolute values of the wavefunctions. The difference between phases is important because it can create interference patterns (as in the double-slit experiment in which two particles create an interference pattern). So the absolute value of the phase is effectively unobservable – the only thing we can observe is the difference between the phases.[21]

So all we can measure is a difference – the absolute value is unobservable. This is therefore very similar to the earlier discussion of electromagnetic potential (voltage). In our discussion of electromagnetic potential, we saw that it was only the difference of the potential that mattered – the absolute value of the potential was irrelevant (in fact, it was unobservable). We have therefore now made a remarkable connection between electromagnetic potential (voltage) and

---

[21] Why is the phase of a wavefunction unobservable? Because quantum mechanics can only tell us the probability of an event occurring, for example, the probability that a particle will be at a certain position. If you read my fourth book you will know that the German physicist Max Born showed that the value of the **square** of the wavefunction represented the probability that the electron is found at a particular location. But when we square the wavefunction, we lose all the phase information (the phase being represented by a complex number). So the phase information is fundamentally unobservable – we can only measure the probabilities.

the phase of the wavefunction of a charged particle: only differences matter – absolute values are unobservable.

This connection is not a coincidence. Instead, this reveals the very deep connection between the phase of the wavefunction of a charged particle (such as an electron) and the absolute value of the electromagnetic potential of the underlying field. As Bruce Schumm says in his book *Deep Down Things*: "The quantum-mechanical requirement of local phase invariance, miraculously, just matches the freedom of choice one has in specifying the electromagnetic potential function for those fields."

Now remember that we considered the unobservable motion of the electromagnetic potential as being motion in a hidden dimension, the hidden fifth dimension in Kaluza's theory. So Felix Klein now made the connection between the varying phase of the wavefunction (motion around a circle) and the motion in Kaluza's hidden fifth dimension. Hence, Klein came to a remarkable conclusion: the fifth dimension must be in the form of a circle. According to Klein, electromagnetism – as experienced in our four dimensions of spacetime – came from motion around this hidden circular fifth dimension.

But can we really make the leap from rotational gauge symmetry of the electromagnetic force to the suggestion that this represents an additional dimension of space? The physicist Juan Maldacena appears dubious: "In physics, we do not know whether this circle is real. We do not know if indeed there is an extra dimension. All we know is that the symmetry is similar to the symmetry we would have if there was an extra dimension. In physics we like to make as few assumptions as possible. An extra dimension is not a necessary assumption, only the symmetry is."[22]

Remember back to the discussion in Chapter Two of how unification can be achieved by "glueing" together two different dimensions into a higher-dimensional space. Well, Maldacena's quote is suggesting that this is a classic example of that approach. The two dimensions being "glued" together is four-dimensional spacetime and the one-dimensional symmetry of the electromagnetic force. Yes, as Maldacena suggests, there is a common element: they both exhibit symmetries. However, does the combined higher-dimensional space represent an actual physical entity, or are we merely creating another "flying octopus"?

Microscopic dimensions

Nothing in our discussion so far has explained why we are unable to detect this supposed circular fifth dimension. We can clearly detect motion in the three dimensions of space, and we can also detect motion in time, which is the fourth dimension (we can detect motion in time simply by using a clock). So how can this circular fifth dimension be hidden? Where is it?

To explain the missing dimension, Klein came up with a bold and – as it turns out – extremely influential idea. Remember, this was 1926, and Klein decided to incorporate some of the ideas from the newly-discovered quantum mechanics. In particular, he used the result discovered by Louis de Broglie which stated that a particle also acts like a wave (this was the complementary result to Einstein's

---

[22] Juan Maldacena, *The symmetry and simplicity of the laws of physics and the Higgs boson*, http://arxiv.org/abs/1410.6753

discovery of the photon: Einstein showed that a wave could act like a particle, de Broglie showed that a particle could act like a wave). If you read my previous book you will know that Niels Bohr used this result of de Broglie to suggest that only a whole number of electron wavelengths could fit in an orbit around the nucleus of an atom:

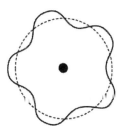

Klein did a very similar thing to Bohr, but instead of suggesting that only a whole number of wavelengths could fit around the nucleus of an atom, Klein suggested that only a whole number of particle wavelengths could wrap around the circular fifth dimension. So how long would those wavelengths be? Well, Klein used de Broglie's formula for the wavelength of a particle:

$$\lambda = \frac{h}{p}$$

where λ is the particle wavelength, p is the momentum of the particle, and h is Planck's constant. Because h is such a small value (6.63×10^{-34} joule seconds) the formula predicts an extremely small value for particle wavelength (which explains why we never normally see particles – or matter – acting like a wave). In fact, because of the small value of h, de Broglie's formula predicts a wavelength for the fifth dimension which is an extraordinarily small distance – close to the Planck length, which is the smallest distance we can meaningfully measure (the Planck length is a thousandth of a millionth of a billionth of a billionth of a billionth of a metre).

So this provided an explanation of why the circular fifth dimension was hidden from our eyes: it was wrapped in incredibly small circles, a smaller distance than any current instrument can probe.

This implies that each point in our four-dimensional spacetime is actually represented by a circle, or loop, in the microscopic fifth dimension. Another way of looking at it is that a one-dimensional line is actually a cylinder (a series of loops). The following diagram shows a section of an apparently one-dimensional line being magnified to reveal it is actually a microscopic cylinder (being formed of a series of loops):

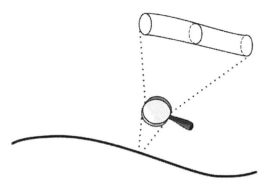

So, if we were sufficiently small, then as well as travelling up and down the line (in one dimension), we could also travel in a circular direction around the cylinder (adding another dimension).

This process – by which higher-dimensions are looped and are too small for us to detect – is called *compactification*.

These two combined theories of Theodor Kaluza and Felix Klein – suggesting that electromagnetism is a result of a microscopic circular fifth dimension of space – is called *Kaluza-Klein theory*.

And, if you are a regular reader of popular science books, you might now be suspecting that this idea of Kaluza-Klein – microscopic loops in higher dimensions – sounds strikingly similar to string theory. In that case, you would be absolutely correct: Kaluza-Klein theory is now recognised as an important precursor to string theory.

6

STRING THEORY

The last thirty years of fundamental physics research has been highly unusual in its nature. It has been dominated by a single theory – *string theory* – which was conceived in the early 1970s. The theory is particularly relevant to this book as it predicts a universe with a large number of hidden compactified dimensions.

In our study of string theory, let us first return to consider an implication of Kaluza-Klein theory. We have seen that the fifth dimension is proposed to be circular and microscopic. We have also seen how the addition of a fifth dimension has the effect of introducing the electromagnetic field in four-dimensional spacetime. But we know that the electromagnetic field is composed of billions of photons (the particles of electromagnetic radiation – such as light). So let us now imagine that the fifth dimension is completely empty of all matter. If that is the case, then it appears that a photon (a field particle) has been produced purely by a loop of fifth-dimensional space. In other words, a particle has been produced purely by a loop in space: a particle purely from geometry!

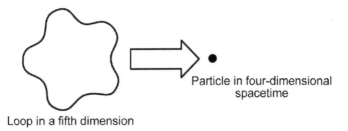

Loop in a fifth dimension

Particle in four-dimensional spacetime

String theory generalises this result. String theory says that a particle is not produced by an empty loop of a higher-dimensional space, but there is actually a physical object which loops around in that higher dimension. Fairly obviously, this is called a *string*, and it gives the name to string theory. The string is not "made" of anything else – it is fundamental.

By introducing a physical object, this introduces the possibility of new structures and modes of behaviour – we are no longer restricted to considering just a simple closed loop. As an example of this, some string theories allow the possibility of *open strings* as well as *closed strings*:

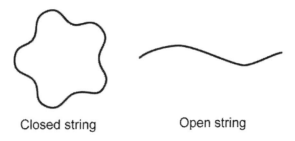

Closed string

Open string

In string theory, therefore, particles are considered as being strings in higher dimensions. Because these higher dimensions are so small, the strings actually appear like microscopic particles. So, according to string theory, elementary particles such as electrons and quarks are not pointlike particles at all but are actually microscopic strings. These strings are so small that they even appear like pointlike particles under the highest magnification currently available: we would need accelerator energies a million billion times more powerful than our current particle accelerators to be able to observe these strings.

There are some theoretical advantages from considering particles as being strings. In order to understand why, we need to consider how particles exist in spacetime. As a particle naturally moves forward in time (time passes), it is clearly able to move in space as well over that period. The following diagram shows both space and time plotted in a diagram – a *spacetime diagram*. Note that only one dimension of space is shown along the vertical axis (it is not possible to plot more than one dimension of space in these diagrams as the other axis represents the time dimension). In the diagram on the left you can see how a particle plots a curved line in a spacetime diagram as it moves in space (on the vertical axis) as time progresses (along the horizontal axis). This traces the position of the particle at each moment in time. For a particle, this line is called a *worldline*:

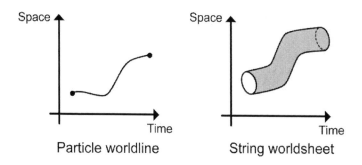

Particle worldline   String worldsheet

Strings, however, are clearly not infinitely small point-particles. We have seen how strings can be looped structures, so as these loops move through time they actually plot a two-dimensional *worldsheet* (as shown on the right in the previous diagram).

So, bearing in mind that a particle plots a worldline whereas a string plots a worldsheet, let us now examine the spacetime diagram for particle interactions. In the following diagram, the diagram on the left shows two particles colliding to produce a third particle (for example, this might be an electron and a positron annihilating to produce a photon):

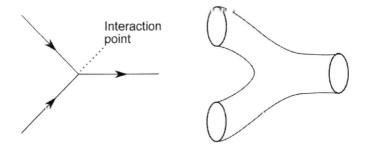

You will see there is a clearly-defined *interaction point* where the two particles meet. Having such a clearly-defined infinitely small interaction point for the two particles introduces problems: unavoidable infinities enter the equations describing this interaction, and infinities must be avoided in any accurate model of the natural world (it is the apparently zero distance between the particles at the interaction point which leads to the infinities).

However, the diagram on the right in the previous diagram shows the same interaction but this time the particles are represented by string loops. So instead of the particles plotting a one-dimensional worldline on the spacetime diagram, the strings now plot two-dimensional worldsheets. As you can see on the diagram, this now avoids there being any clearly-defined interaction point – the strings have the effect of spreading the interaction. The zero-distance problem – and hence the undesirable infinities in the equations – can be avoided.

By taking slices through the worldsheet at different instances of time we can see how the worldsheet represents two string loops (two particles) interacting to produce a third particle. In the following diagram, you can see the interactions of the string loops underneath the worldsheet. You can see that two string loops (coming in from the left) representing two different particles interact to produce a single string loop (a single particle). Notice how smooth the interaction has become:

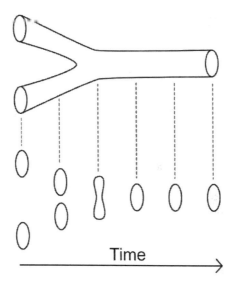

Time

The music of strings

One of the most surprising features of string theory is that the strings are modelled very much as if they were everyday strings – in fact, just like a guitar string. Yes, string theory really is a theory of physical strings! And, just like a guitar string, a string in string theory can oscillate at a certain frequency. If we imagine a guitar string fixed at both ends, we can see that only particular frequencies are possible.

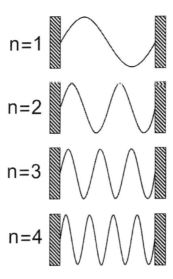

Considering the diagram above, you can see that it is possible for the two ends of the guitar string to loop back round and connect to each other, thus forming a closed loop with a discrete (integer) number of wavelengths allowed:

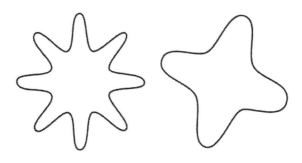

(This might remind you of the whole number of electron wavelengths fitting in orbit around the Bohr model of an atom).

Just as with a guitar string, we can think of these different *modes* of vibration as representing different musical "notes". So what do these different vibrational modes represent? Well, intuitively, we can think of the shorter wavelengths as representing more energetic motion, so the higher the wavelength, the greater the energy of the string. You might also be reminded of Planck's result which linked the energy of a photon to the frequency of the electromagnetic radiation carried by that photon in which, again, higher frequencies (shorter wavelengths) mean more energy. And, through $E=mc^2$, if a string has more energy then that means it must have more mass. So the pattern of vibrations of a string determines the mass of that string and, therefore, the mass of the elementary particle which the string represents.

This is a major departure from the usual way in which we think about the mass of particles. We usually think of particles as being fundamentally different, made of different "stuff", so to speak. However, if string theory is correct, then all particles are made of exactly the same thing: a string, and the only thing which differentiates the particles is the particular mode of vibration of that string. The mass of the particle is determined by the vibration of the string.

Let us now consider the lowest mode of string vibration, with the lowest possible frequency and therefore the lowest possible mass:

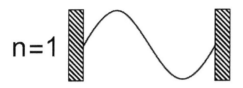

If the value $n = 1$ is entered into the formula for string mass, then the formula gives a value of zero, which means the associated string would have no mass. This appears to indicate the presence of an unknown massless particle, and this was initially considered to be a failed prediction of string theory. However, in 1974, Joel Scherk and John Schwarz suggested that this massless particle might well be the *graviton*, the particle which is proposed to carry the force of gravity (the graviton would have to be massless to transmit the force of gravity across the entire universe via gravitational waves).

We have already seen that string theory can predict the photon, the particle which transmits the electromagnetic force. String theory now appeared to predict gravity, another of the fundamental forces. At a stroke, string theory changed from a theory with limited interest to potentially being able to unify the four fundamental forces. From this point on, string theory emerged as the leading contender to be a potential "theory of everything".

The best-kept secret in mathematics

I hope you have enjoyed this gentle introduction to string theory. Unfortunately, nothing so far in this chapter on string theory has been directly relevant to this book and its central theme of the number of spatial dimensions, which is what we will consider in this section.

In this discussion of the basic principles of string theory I hope the theory has appeared reasonable, with nothing remotely astonishing or bizarre. This has been quite a deliberate approach on my part, because things are about to get seriously weird!

We are about to derive the number of spacetime dimensions predicted by string theory.

If someone was to tell you that the sum of all the numbers was $-\frac{1}{12}$ I suspect you might imagine the person was mildly crazy. Well, this is actually an established mathematical fact: the sum of all the numbers is minus one twelfth (or, in decimals, -0.083). Let me be more precise about what, specifically, is being stated here. We are actually referring to the sum total of all the **natural** numbers. We discussed the natural numbers in Chapter One: they are the numbers used for counting: 1, 2, 3, etc. So what is actually being stated here is that the infinite sum of 1+2+3+4 … comes to minus one twelfth.

Let's write that out:

$$1 + 2 + 3 + 4 + 5 + \ldots \infty = -\frac{1}{12}$$

We can express this infinite sum in mathematical notation by the following mathematical expression, which is taken directly from Joseph Polchinski's popular and authoritative textbook on string theory which is called, simply, *String Theory*:

$$\sum_{n=1}^{\infty} n \rightarrow -\frac{1}{12}$$

On the left hand side of the expression, you will see the Greek letter sigma (Σ) which represents a sum. You will see that in this case it represents a sum of the variable n as n takes all the values from 1 to infinity (so this is just a mathematical way of representing the infinite series $1+2+3+4$ ...). The right hand side of the expression shows that this series will indeed eventually take the value minus one twelfth.

You might not believe this is true: how can an infinite sum of positive numbers ever result in a finite negative value? I'll admit, this is a difficult result to understand, but I can only repeat that this is a generally-accepted important result in mathematics and physics, and is absolutely central to string theory. It is central to string theory because this result is used by string theory to predict the number of spatial dimensions.

We will now derive this result. I have to say that this derivation can also be found in an excellent video on YouTube in which two physicists from the University of Nottingham explain it in a very entertaining manner. I can highly recommend you watch the video:

http://tinyurl.com/sumofallnumbers

The video went viral and currently has four million views on YouTube, which I am sure is a record for a video about string theory. Well done to those physicists.

There is also a *New York Times* article which considers the result: **http://tinyurl.com/timesinfinity**

In that *New York Times* article, Edward Frenkel, a mathematics professor from Berkeley, says: "This calculation is one of the best-kept secrets in mathematics. No one on the outside knows about it."

I would suggest that four million YouTube viewers might now disagree with the professor.

OK, so now let us proceed with this amazing derivation (as I said, the YouTube video of the Nottingham physicists is perhaps a clearer version of this). Let us start by writing out that infinite sum of natural numbers, and let us call the result of the infinite sum S:

$$S = 1 + 2 + 3 + 4 + 5 + 6 + ...$$

Remember, we want to calculate the value of S.

Now let us consider a second infinite sum. Let us define the sequence S1 as:

$$S1 = 1 - 1 + 1 - 1 + 1 - 1 + 1 ...$$

Note that the signs alternate from minus to plus on each number. What is the total value of S1? Well, you will see that if we stop the sequence after a +1 then the sequence S1 will have the value +1. Alternatively, if we stop the sequence after a -1 then the sequence S1 will have the value 0. So what possible single numeric value can we give the sequence if the sequence extends to infinity? The only value which makes any sense is an average of +1 and 0, which means the value of the infinite sum S1 must be 0.5.

Now let us consider a third infinite sum, S2:

$$S2 = 1 - 2 + 3 - 4 + 5 - 6 + 7 ...$$

Let us now calculate the value of the sum S2 multiplied by two. We can do this by adding S2 with itself, but we will shift this second version of S2 by one place to the right:

```
S2=1-2+3-4+5-6+7...
S2=   1-2+3-4+5-6...  +
      1-1+1-1+1-1+1...
```

If you consider the sums of each number vertically you can see that the result of this addition is the infinite sum of $1-1+1-1+1-1+1$ which is just the sequence S1. And we already know that S1 has the value 0.5. Therefore:

```
2×S2=0.5
```

which means that S2 has the value 0.25.

Now let us consider what we get when we subtract the sequence S2 from sequence S (in the following subtraction, remember that subtracting a negative number turns it into a positive number):

```
S =1+2+3+4+5+6...
S2=1-2+3-4+5-6...  -
      4+0+8+0+12...
```

so we get the infinite sum $4+8+12+16...$ which is the series S multiplied by four. Therefore:

```
S-S2=4S
```

We already know that S2 has the value 0.25, so this means:

```
S-0.25=4S
```

Subtracting S from both sides gives:

```
-0.25=3S
```

so therefore:

$$S= -\frac{1}{12}$$

and that's it! The sum of the infinite sequence S, which is 1+2+3+4 ... is minus one twelfth!

You might find this hard to believe, but I can assure you that this is an established and accepted result in mathematics, and it was rigorously proved by the great mathematician Leonhard Euler in 1749. The result is also used in physics in a formula which is completely unconnected with string theory and gives completely accurate experimentally-verified predictions about the behaviour of Nature.[23] So this is a correct result!

The recently-released film *The Man Who Knew Infinity* tells the story of the Indian mathematician Srinivasa Ramanujan. Ramanujan was born in 1887. He came from a background of extreme poverty and had no formal training in mathematics, but he had a tremendous natural talent and instinct for the subject, with a particular talent for constructing infinite series. Ramanujan sent some of his

---

[23] It is used in the calculation of the *Casimir effect*. If two plates are placed just a few nanometres apart in a vacuum, there will be a small attractive force between the plates due to the vacuum energy of the electromagnetic field. The -1/12 value explains why summing over all the possible vacuum energy states does not result in an infinite value. The minus sign explains why the Casimir force is attractive.

original work to the mathematician G.H. Hardy at Cambridge University. Hardy was amazed at Ramanujan's skill and original methods. Hardy invited Ramanujan to study at Cambridge, and Ramanujan eventually became a Fellow of the Royal Society.

Ramanujan's story is relevant to this chapter because he also derived the result that the infinite sum of the natural numbers is equal to minus one twelfth. Ramanujan's method was very similar to the method described earlier and shown on YouTube. After calculating this amazing result, Ramanujan wrote to Hardy:

Dear Sir, I am very much gratified on perusing your letter of the 8th February 1913. I was expecting a reply from you similar to the one which a mathematics professor at London wrote. I told him that the sum of an infinite number of terms of the series 1 + 2 + 3 + 4 ... = -1/12 under my theory. If I tell you this you will at once point out to me the lunatic asylum as my goal.

Ramanujan's derivation is shown below in his own handwriting:

I would suggest that this is one of the most remarkable results in the whole of science.

So what has this got to do with string theory? Well, we have already seen that there are an effectively infinite number of modes of oscillation for a string ($n = 1$, $n=2$, etc.). So the total number of modes in which a string can oscillate will be an infinite sum of all these modes:

$$\sum_{n=1}^{\infty} n$$

So the infinite sum of all the natural numbers appears in the equation giving us the number of modes in which a string can oscillate. We now see that our earlier apparently crazy piece of infinite arithmetic has relevance to string theory. What is more, our crazy piece of arithmetic has tamed the infinities! As was explained earlier, the mass of a string (and, therefore, its energy) depends on all the modes of vibration. If the sum of all the modes of vibration had been an infinite number, then the energy of a string would be infinite and string theory could not possibly be a correct theory of reality (we do not see infinities in the physical world). However, our apparently crazy arithmetic has produced a finite result, so string theory can be a *finite* theory. This is a vital feature of any theory.

For the next part, we need to consider the number of directions in which a string can oscillate. This is described by Joseph Conlon in his book *Why String Theory?*:

> A string can be plucked, and can oscillate, in every direction transverse to its length. A string living in two spatial dimensions has one direction it can oscillate in; a string in three spatial dimensions has two directions it can oscillate in; a string in twenty-five spatial dimensions has twenty-four dimensions it can oscillate in.

On that basis, a string in three dimensions of space (four dimensions of spacetime) can oscillate in two perpendicular directions – transverse to its length – and this is shown in the following diagram:

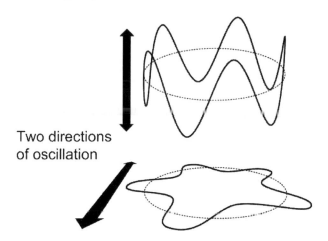

Two directions of oscillation

This reasoning implies that, in D dimensions of spacetime, a string is able to oscillate in $(D\text{-}2)$ directions.

To calculate the total energy of a string, we have to consider the infinite sum of all the modes in which a string can vibrate, and multiply that sum by the number of directions in which the string can vibrate. Hence, the sum of all energies of the string is given by:

$$(D-2)\sum_{n=1}^{\infty} n$$

Using this result, we are now in a position to calculate the number of dimensions, D, predicted by string theory. Without going into details, the actual equation we need to solve to find D is:[24]

$$\frac{1}{2}(D-2)\sum_{n=1}^{\infty} n = -1$$

We have just shown that the infinite sum in this equation is equal to minus one twelfth. So let us replace the sigma term in the equation with the value minus one twelfth, which results in:

$$\frac{1}{2}(D-2)\times -\frac{1}{12} = -1$$

Multiplying both sides of the equation by -24 gives:

$$D-2 = 24$$

which means $D=26$. Therefore, string theory predicts 26 dimensions of spacetime! In other words, string theory predicts one dimension of time, and 25 dimensions of space.

I think the immediate response of most people on discovering a theory predicts 25 dimensions of space would be a feeling of disappointment shortly followed by a rejection of the theory for being incorrect. As Lee Smolin

---

[24] The expression on the left-hand side actually represents the sum of the lowest energies of each oscillator mode (the so-called *zero-point energies*). The ½ term comes from the standard result for the lowest energy of a quantized harmonic oscillator (a vibrating string is a harmonic oscillator). The -1 term on the right-hand side represents the fact that the ground state energy is not zero, but has a small negative value called the *Casimir energy* (the value is the negative of the $n=1$ lowest-energy state).

says in his book *The Trouble with Physics*: "The world does not appear to have twenty-five dimensions of space. Why it is that the theory was not just abandoned then and there is one of the great mysteries of science." However, the theory was most certainly not abandoned. With string theory, I think the phrase "putting a positive spin on things" takes on a whole new meaning. A direct quote from one of my string theory textbooks provides an example of how this apparently negative outcome is frequently presented as a positive feature: "Since the dimension of spacetime is uniquely selected by the requirement of consistency, we can say that string theory predicts the dimension of spacetime!" [25]

Well, I suppose that predicting a finite integer number of spacetime dimensions must be considered some kind of an achievement. It is just a shame that it is the wrong finite integer number of spacetime dimensions.

---

[25] Chapter 12 of *A First Course in String Theory* by Barton Zwiebach.

The hidden dimensions of string theory

So far we have seen that string theory is able to predict the existence of the photon and the graviton, in other words string theory can predict the particles which transmit the forces. If you read my previous book, you will know that these intangible particles are called *bosons*. For this reason, the version of string theory we have considered so far is called *bosonic string theory*.

Bosonic string theory was the original version of the theory developed in the early 1970s and might be considered to be the core of string theory. However, if you read my previous book you will know that all particles are divided into one of two groups: they are either bosons, or they are *fermions*. Whereas bosons are the intangible particles (which form a light beam, for example), fermions are the particles which make matter. So fermions are the particles which form a chair, a car, a human, for example. And the trouble with bosonic string theory is that it cannot be used to produce fermions. Hence, bosonic string theory on its own cannot be a realistic theory.

A solution to this problem emerged in the 1970s in the form of *supersymmetry*. Supersymmetry predicts a symmetry in which bosons can be exchanged for fermions. So incorporating supersymmetry into string theory produced a complete theory which could now predict the existence of fermions.

String theory which incorporates supersymmetry is called *superstring theory*. One attractive side-effect of superstring theory is that the predicted number of spacetime dimensions was reduced to ten. It is clearly still not correct, but it might be regarded as a step in the right direction.

Supersymmetry also predicts the existence of a huge range of additional undiscovered particles called *superpartners*. By incorporating supersymmetry, superstring theory has made itself something of a hostage-to-fortune to the discovery of these additional particles. If the superpartner particles are not discovered, then it would deal a fatal blow to superstring theory. So far, the Large Hadron Collider at CERN has not discovered any supersymmetric particles. This is bad news for supersymmetry and, therefore, bad news for superstring theory.

By the late 90s, string theory was not in the best of shape having splintered into five variations. However, in 1998, Ed Witten (the effective leader of the superstring theory field) announced that it might be possible to unify these five theories into a single theory (which he called *M-theory*). Unfortunately, this unification was achieved by the usual method of introducing an additional dimension of space. String theorists greeted the M-theory proposal with great excitement – despite the newly-predicted eleven dimensions of spacetime being even more inaccurate than the previous ten-dimensional prediction.

Whichever variation of string theory you consider, one feature is clear: the theory will predict considerably more than four spacetime dimensions. The problem then arises: where are these extra dimensions, and why do we not see them? The proposed solution is based on the Kaluza-Klein idea that the extra dimensions are compactified – too small to detect. A particular structure – called the *Calabi-Yau space* – has been suggested as a way in which six spatial dimensions might be compactified (much like the single spatial dimension is compactified into a circle in Kaluza-Klein theory). The following diagram shows a projection (in two-dimensions, obviously) of the six-dimensional Calabi-Yau space:

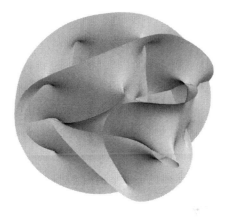

Unfortunately, it appears there are many tens of thousands of different Calabi-Yau spaces, and string theory does not indicate which is the correct one. What is more, each different Calabi-Yau space – each different folding of the extra dimensions – would result in different laws of physics, and a different set of elementary particles. If string theory is unable to select one particular Calabi-Yau space then it would appear that string theory is unable to predict a unique set of laws of Nature.

It has even been suggested that there might be a vast "landscape" of possible universes with different Calabi-Yau foldings in each universe. This represents another form of the anthropic argument in which virtually any outcome is possible. As a result, sting theory has been criticised as being untestable and potentially unscientific.

For these reasons, it appears that string theory research is moving away from an attempt to find a fundamental theory of quantum gravity. Instead, the field appears to be splintering, with some researchers more interested in string theory purely because of the new mathematical techniques that have been developed.

7

WHY THREE DIMENSIONS?

OK, this is the last chapter, so let's have a bit of fun.

If you have read my previous books, you will know I like to include some of my own original ideas in the later stages of my books, and this book is no exception. So far, we have considered several of the main hypotheses as to why there are three spatial dimensions, and in this chapter I will present my own idea. So bear in mind that the material presented in this chapter should be considered speculative (however, all the material in this book – with the exception of the chapter on general relativity – should be considered speculative as absolutely no one knows the real reason why there are three dimensions of space).

The solution which will be presented here is particularly appealing because it includes principles from almost every area of fundamental physics: relativity, quantum mechanics, particle physics, and thermodynamics. It will also include several of the "fundamental principles" which have been considered in my previous books.

I had several guidelines in mind when I developed this theory:

- The theory would have nothing to do with anthropic reasoning. Therefore, it could include no reliance on unseen parallel universes. Instead, a unique solution would have to emerge from clear logical reasoning. As Joseph Conlon says about anthropic theories in his book *Why String Theory?*: "The argument lacks the redeeming precision of cut-and-dried mathematical argument." We will be searching precisely for a cut-and-dried argument, and we will even be ending up with a simple equation which gives us the answer we seek: the number three.

- The theory would clearly and unambiguously have to predict three spatial dimensions. Not ten dimensions – three dimensions. This is in line with the naturalness argument (considered in Chapter One) which suggests a number of dimensions close to the value of one, and it is also in agreement with every observation which has ever involved counting the number of spatial dimensions.

- As explained at the end of Chapter Two, we should not expect to find three-dimensional space as a pre-existing absolute background. Instead, we should expect three-dimensional space to **emerge** from the properties of the fundamental particles which compose the universe. This principle was described by Einstein in the *New York Times* in 1919: "Till now it was believed that time and space existed by themselves, even if there was nothing – no Sun, no Earth, no stars – while now we know that time and space are not the vessel for the universe, but could not exist at all if there were no contents, namely no Sun, no Earth, and other celestial bodies." Einstein repeated this insight that time and space emerged from the behaviour of matter in the *New York Times* in 1921: "Up to this time the conceptions of time of time and space have been such that if everything in the universe were taken away, if there were nothing left, there would still be left to man time and space." Einstein had realised that if there is no matter, there is no space. To find our solution, we need to consider the behaviour of matter.

- The theory would have to be simple. We are dealing with the most fundamental of physical entities: the number of dimensions of space. It is a firm belief of all my books that, as we approach the fundamental level, our theories should get simpler — not more complicated.

The proposed solution presented in this chapter will be composed of five simple parts:

1. A bit of relativity.
2. A bit of thermodynamics.
3. A bit of particle physics.
4. A bit of quantum mechanics.
5. Why three dimensions?

OK, let's get started …

Part 1: A bit of relativity

As explained earlier in this book, Hermann Minkowski realised that special relativity suggested that time and space should be combined into a higher-dimensional combined spacetime. And it is this spacetime model which provides us with a simpler way of looking at special relativity. This simplification is described by Brian Greene in his excellent book *The Fabric of the Cosmos*: "The combined speed of any object's motion through space and its motion through time is always precisely equal to the speed of light." In other words, it can be considered that all objects travel through spacetime at the speed of light. As mentioned in Chapter Four earlier in this book, even an object which is stationary, i.e., is not moving in space, is clearly still moving through time (a watch, for example, clearly shows the passage of time). For the complete details of why this speed through spacetime is necessarily the speed of light, see my third book.

This principle allows us to understand special relativity in a simpler form. As Brian Cox and Jeff Forshaw explain in their book *Why Does $E = mc^2$*: "This newfound way of thinking about how things move through spacetime can help us get a different handle on why moving clocks run slow. In this spacetime way of thinking, a moving clock uses up some of its fixed quota of spacetime speed because of its motion through space and that leaves less for its motion through time."

This interpretation of special relativity provides us with a simpler way of understanding the time-travelling astronaut. An astronaut who flies away from the Earth at nearly the speed of light before returning to the Earth will have travelled a greater distance through space than his friend who stayed on Earth. However, because the astronaut will have travelled a greater distance through space, he will have travelled a smaller distance through time (as everything travels at the same speed in spacetime). In other words, he will have aged less.

So special relativity provides us with the first principle we need on our quest to understand why there are three dimensions of space: everything travels through spacetime at the speed of light.

Part 2: A bit of thermodynamics

The nature of time still remains something of a mystery to us. This is because – certainly down to the atomic level – all physical processes appear to be reversible with time.[26] For an example from Newtonian mechanics, consider a movie of a moving ball colliding with a stationary ball. The first ball would stop and the second ball would move off at speed. If the movie was played backwards, the events would still make sense according to the laws of physics. This time, though, the second ball would come in reverse, strike the stationary first ball, and the first ball would then move off in reverse. Everything would happen perfectly in reverse, and it would look as though it was happening in the forward time direction.

---

[26] At the level of individual particles, the interactions of the weak force do not appear to be completely symmetric with respect to time. However, according to *CPT symmetry*, if the charge of a particle (C) and the parity of a particle (P) and time (T) are all reversed then all physical interactions are indeed reversible – even interactions due to the weak force.

So if the laws of physics are time-symmetrical, why do so many processes exhibit a so-called *arrow of time* in the forward time direction? For example, we might see an egg breaking (in the forward time direction), but we never see a broken egg reforming itself – so clearly that is not a process which is time-symmetrical. The answer as to why there is an arrow of time comes from the increase in *entropy*, which can be regarded as the amount of disorder in a system. It is known that the amount of entropy in a system will always tend to increase with time, in other words a system will become more disordered over time. Hence, we see cars rusting (their molecules becoming more disordered), but we do not see rusting cars becoming new again. This principle that the entropy of a closed system increases with time is called the *second law of thermodynamics*.

But entropy is a statistical property which can only be measured from a group of particles. It therefore makes no sense to talk about the entropy of a single particle – a single elementary particle has no internal parts, so it can never get disordered, or fall to bits over time. In a similar fashion, it makes no sense to talk about the temperature or pressure of a single particle (temperature and pressure are also statistical properties of a group of particles). As Joseph Conlon says in his book *Why String Theory?*: "The concept of temperature requires many particles."

So – if this is the case – then how does a single elementary particle experience time? If it is not subject to increasing entropy then it is not subject to the usual arrow of time. However, we know that elementary particles do experience time because of experimental observations. For example, when protons from space hit the upper atmosphere of the Earth, they produce *muons*, which are a heavier form of electron. Because the muons are heavy, that means they rapidly decay to lighter particles: the average lifetime of a muon is 2.2 microseconds. In that very short period of time, even muons travelling at close to the speed of light will only travel 660 metres before they decay. This means very few muons should actually reach the surface of the Earth. However, in practice, many more muons reach the Earth's surface. This is because of time dilation predicted by special relativity: the particles are travelling at close to the speed of light, so time passes slower for the muons. Hence the muons "age slower" and it takes them longer to decay.

It is also well known from particle accelerator experiments that elementary particles take longer to decay when they are moving close to the speed of light. This was described by Luboš Motl in a blog posting entitled *The world as seen by the LHC protons*: "All the processes occurring 'inside' the moving particle are slowed down due to time dilation. That's true for the 'ageing process' of the particle, too."[27]

But if an elementary particle has no internal structure, it can have no internal clock. And if it has no internal parts then it has no entropy and is therefore not subject to the usual entropy-based arrow of time. So how do elementary particles experience time? How do elementary particles "age"? This is a very fundamental question.

We will shortly be returning to this question as we shall see it possibly holds the key as to why there are three dimensions of space.

Bear with me … we are getting there!

---

[27] http://tinyurl.com/lhcprotons

Part 3: A bit of particle physics

The world of particle physics can seem a very confusing place. However, if you read my previous book you will know that all particles can be placed in one of two simple categories: a particle is either a *fermion* or it is a *boson*.

Fermions resist being in the same state as other fermions. It is this tendency which causes fermions to be the particles which form atoms. Electrons are fermions, and they orbit the atomic nucleus (composed of quarks, which are also fermions) in clearly-defined shells. The clearly-defined electron shells prevent two electrons from being in identical states.

Because fermions resist being in the same state, an object composed of fermions will resist our touch, feeling solid. Hence, fermions are the particles which form matter: trees, cars, gases, tables, and humans.

The other category of particles are the bosons. A photon (a particle of light) is an example of a boson. In contrast to the behaviour of fermions, bosons like to be in the same state. Hence, billions of photons will congregate to produce a beam of light. Because photons do not mind being in the same state, bosons feel intangible: a light beam does not resist our touch. Hence, bosons are not suitable particles for forming matter.

This distinction between fermions and bosons is vital for our quest to determine why there are three dimensions of space. This is because of our human bias when considering questions about space (considered in Chapter Two). We are made of fermions. The objects we manipulate are made of fermions. The objects we have to avoid when we navigate space are made of fermions (we generally don't have to avoid a beam of light). Atoms are made of fermions, and galaxies are made of fermions. In fact, it could be said that we define space in terms of fermionic matter.

This is important for our quest to understand why there are three dimensions of space. To see why, let us ask the hypothetical question: how many dimensions would we experience if we were made of light? Special relativity tells us that an object ages slower as its speed approaches the speed of light. An object made of fermions has mass, and can never reach the speed of light. However, light is composed of massless bosons and is therefore able to travel at … well … the speed of light. Light, therefore, does not experience time. Harry McLaughlin, one of the top contributors on the *Quora* website, describes the experience of light quite beautifully:

> *Imagine watching a sunset. A photon is emitted from an atom in the solar photosphere and journeys across space until it reaches your eye and is absorbed by a molecule in your retina. From your point-of-view, this photon had a lifespan of about eight minutes.*
>
> *However, for the photon, its own emission and absorption is a single event. There are no defined notions such as "happening", "process", or "experience".*
>
> *For a photon, time is non-existent and meaningless.*

With this in mind, let us return to our hypothetical question: how many dimensions would we experience if we were made of light? From the previous discussion it is clear that if we were made of light we would not even be asking the question "Why does the universe have four spacetime dimensions?" because we would not experience four spacetime dimensions. Just like light, we would not experience the time dimension. In other words, we would have lost a dimension!

And I think this is a vital point: the number of dimensions of the universe depends on your viewpoint. Is your universe defined by bosons or fermions? This makes it very clear that – to repeat a point made in Chapter Two – the dimensions emerge from the behaviour of matter in the universe. The spatial dimensions do not form a pre-existing system of axes. Newton's idea of absolute space has been refuted. There is no absolute "box" of spatial axes "outside the universe" which would remain if all the matter in the universe was removed. Instead, the dimensions come from the behaviour of that matter: the dimensions emerge from the behaviour of particles.

So, if we truly want to understand why there are three spatial dimensions, we need to consider the behaviour of particles. Specifically, we need to consider the behaviour of fermions, because that is how we define our space.

Part 4: A bit of quantum mechanics

In the first two decades of the 20th century, physics was rocked by the two major developments of quantum mechanics and relativity. However, the incompatibility of these two developments posed (and still does pose) a major challenge for physicists.

If you read my previous book, you will know that in 1928 the British physicist Paul Dirac made a major breakthrough by successfully combining quantum mechanics with special relativity for the first time. In his famous Dirac equation, he managed to accurately describe the quantum mechanical behaviour of an electron which incorporated the rules of special relativity. He achieved this by using a *matrix*. A matrix is a rectangular array of numbers which can be used to transform one set of coordinates into another set of coordinates (for example, a rotation of an object can be described by a matrix). So, in his equation, Dirac used a matrix to transform spatial coordinates into time coordinates – according to the rules of special relativity. This resulted in a description of how the electron behaved in combined spacetime.[28]

So, what did the Dirac equation reveal about the electron? Amazingly, it revealed the electron had to spin. It was already known that an electron possessed the property of spin, but for the Dirac equation to predict the necessity for spin was an extraordinary achievement.

How are we to interpret this necessity for electron spin? Well, perhaps we should not be too surprised. Paul Dirac introduced the rules of special relativity into quantum mechanics, and we know from our earlier discussion that special relativity has a simple interpretation: everything travels at the speed of light in spacetime. If we consider a stationary electron, how can it possibly continue to move through spacetime at the speed of light? The simple answer is: it has to spin. Remember in our earlier discussion of entropy, the question was asked how can an elementary particle move through time? An elementary particle has no internal parts, so increasing entropy cannot explain the arrow of time for a single particle. Instead, perhaps we can consider the necessity for a particle to spin as representing the only way a stationary elementary particle can continue to travel through spacetime. How does a stationary elementary particle travel through time? The only way it can: it spins.[29]

---

[28] The resulting description of the electron is said to be *Lorentz invariant*, meaning that the description of the electron does not depend on how the observer is moving, i.e., the description conforms to the rules of special relativity.

But if particle spin represents a particle's speed through spacetime, at what speed should we expect it to spin? You might be able to guess: we would expect it to spin at the speed of light – the same speed that everything travels through spacetime. This was suggested by Paul Dirac in his Nobel Prize-winning lecture in 1933:

> *It is found that an electron, which seems to us to be moving very slowly, must actually have a very high frequency oscillatory motion of small amplitude superposed on the regular motion which appears to us. As a result of this oscillatory motion, **the velocity of the electron at any time equals the velocity of light.** This is a prediction which cannot be directly verified by experiment, since the frequency of the oscillatory motion is so high and its amplitude is so small.*

However, even though it is impossible to experimentally measure the electron spinning at the speed of light, it is possible to perform a simple calculation to show that this is likely to be the case (the calculation is included in the Appendix).

---

[29] In accordance with the energy-momentum relation in special relativity, translational momentum in the space dimension transforms into angular momentum – spin – in the time dimension (if you are protesting that the intrinsic spin of a particle does not represent actual rotation, please refer to my previous book in which it is argued that it does, indeed, represent actual physical rotation).

But, really, the details of this are irrelevant. The only two things we need to know are that our reality is defined in terms of fermions, and that a fermion must spin. Those two facts alone are enough to determine that space must have three dimensions – as we shall now see …

Part 5: Why three dimensions?

Congratulations – we have made it! This is the last step on our quest to discover why space has three dimensions.

On the basis of what we have just discovered, the solution is very simple. Because of our human bias, our space is defined in terms of fermions. A fermion with mass, travelling slower than the speed of light in the x direction, needs to spin. Spin requires a plane in two dimensions, so an additional two dimensions – y and z, shown below – perpendicular to the direction of motion are required:

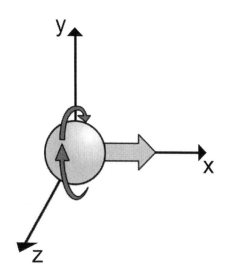

So three dimensions arise from the fermion's need to satisfy the rules of special relativity. And, as we define our reality in terms of fermions, our space therefore must have three dimensions, three degrees of freedom.

Referring back to Chapter One, the idea of three being a "number of sufficient magnitude" was introduced. We can now see how that principle is applicable: one or two dimensions of space would not be enough as the fermion would not be able to spin. Three dimensions is the smallest number of dimensions which allows spin perpendicular to motion, thus allowing the fermion to satisfy special relativity.

But, you might protest, this still does not specifically identify three dimensions of space. OK, one or two dimensions would not be enough, but surely this would still be possible in four dimensions of space, or five or six or any other large number of dimensions? Well, rather wonderfully, it turns out that three dimensional space has a unique characteristic in this regard (as described in the discussion of three-dimensional geometry in Chapter Two, it is often the case that spaces with low number of dimensions can possess unique characteristics which are not shared by higher-dimensional space). John Barrow explains this unique characteristic in *Dimensionality*: "There is one simple geometrical property unique to three dimensions that plays an important role in physics: universes with three spatial dimensions possess a unique correspondence between rotational and translational degrees of freedom. Both are defined by only three components."

This is going to require some explanation.

Imagine a two-dimensional plane: a flat desert. And on that flat desert there is a car. The car is free to drive anywhere on the desert, so the car can move anywhere across that two-dimensional space. In other words, the car has two translational degrees of freedom. But how free is the car to rotate? The car rotates around the vertical axis when it turns left and right – when the driver turns the steering wheel. So the car has only one rotational degree of freedom (hence, the car has only one steering wheel – a steering wheel being required for each rotational degree of freedom).

To sum up, in two-dimensional space there are two degrees of translational freedom, but only one degree of rotational freedom. So, in two dimensions, the number of translational degrees of freedom is not equal to the number of rotational degrees of freedom.

Now let us consider the situation in three-dimensional space.

Unlike a car – which is restricted to movement in two dimensions – an aeroplane is free to move in a straight line in any direction in three dimensional space: forward/back, up/down, left/right. So an aeroplane has three translational degrees of freedom. But how free is the aeroplane to rotate? It turns out that there are also three rotational degrees of rotation available to the aeroplane, and these are called *roll*, *pitch*, and *yaw*.

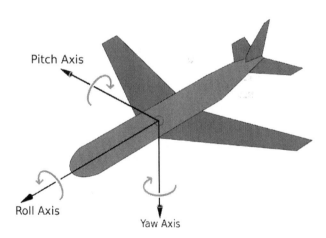

So an aeroplane has three rotational degrees of freedom. For this reason, an aeroplane effectively needs three steering wheels – one for each rotational degree of freedom – and these are incorporated into a joystick:

Just to make it clear, a car or boat is able to move in two dimensions, but only needs one steering wheel (it is only able to rotate around one axis). An aeroplane can move in three dimensions and needs three steering wheels (it is able to rotate around three axes). As John Barrow said, in three dimensions – and three dimensions only – the number of rotational degrees of freedom is equal to the number of translational degrees of freedom.

There is a formula which gives the number of "steering wheels" (number of rotational degrees of freedom) in n dimensions of space, and that formula is:

$$\tfrac{1}{2}n(n-1)$$

Try calculating the result of this formula for various values of n, such as $n=1$ or $n=2$.

- For $n=1$ (one-dimensional space – a line) you will find the formula gives the value 0. This is because you cannot rotate in one-dimensional space, you can only move forwards and backwards along the line, so no "steering wheels" are required. So, in the case of one-dimensional space, the number of steering wheels, 0, is clearly not equal to 1, the number of dimensions.

- For $n=2$ (two-dimensional space – a plane) you will find the formula gives the value 1. This is the case of the car or the boat which can travel in two dimensions but are only able to rotate about a single axis – so only have one steering wheel. So, once again, the number of steering wheels is clearly not equal to the dimensionality of the space.

- For $n=4$ you will find the formula gives the value 6, which is clearly not equal to 4.

However, for $n=3$ you will find the formula gives the value 3. So three steering wheels are required to navigate three-dimensional space. **And, crucially, it can be shown that this is only true for $n=3$.** No other number of dimensions will satisfy this formula.

Effectively, we are solving the magic equation:

$$n = \tfrac{1}{2} n(n-1)$$

You might like to use a bit of high-school algebra to calculate the value of *n* from that equation yourself. You will find that putting $n=3$ satisfies the equation, which is clearly the correct answer for the number of spatial dimensions.[30] This makes it clear what an important role is played by spin (rotation) in defining the number of spatial dimensions.

So that is my proposal for why there are three dimensions of space.

To recap, we define our space in terms of fermions, which are the matter particles such as electrons. A fermion has to comply with the rules of special relativity (is said to be *Lorentz invariant*) and hence requires a plane of rotation in order to spin. But only in three-dimensional space is a single plane of rotation uniquely specified which is perpendicular to the particle's direction of motion.

And that is why I believe three is the magic number.

Thank you.

---

[30] $n=0$ is the other – not very interesting – solution.

APPENDIX

This appendix presents a simple calculation to show that a particle spins at the speed of light (a similar calculation was performed in the early days of quantum mechanics).

In my previous book, it was stated that elementary particles are considered as being infinitely-small pointlike particles. But if the radius of a particle is truly zero, then how can we measure the speed at which a point on the surface of the particle spins? Well, quantum mechanics introduces an element of uncertainty into all measurements due to the Heisenberg uncertainty principle. Because of wave/particle duality, every particle can be thought of as having a wavelength as well. In the following calculation, the radius of the particle will be given by this wavelength, the *Compton wavelength*, which is equal to:

$$r = \frac{h}{mc}$$

where m is the mass of the particle, c is the speed of light, and h is Planck's constant. We will be modelling a particle by a spinning disk which has an angular momentum, L, equal to:

$$L = \frac{mvr}{2}$$

where v is the velocity with which the particle is spinning.

Replacing the value of r in this equation with the value of the Compton wavelength gives:

$$L = \frac{mv}{2}\left(\frac{h}{mc}\right)$$

Replacing the value of angular momentum with the spin angular momentum, $h/2$, gives:

$$\frac{h}{2} = \frac{mv}{2}\left(\frac{h}{mc}\right)$$

You will see that the h, and the m, and the 2 all cancel, leaving us with $v=c$. In other words, the particle spins at the speed of light.

FURTHER READING

A Beautiful Question by Frank Wilczek
A lavishly-illustrated book, containing some very deep and interesting ideas. Includes sections on the Platonic solids, and beauty in Nature.

The Constants of Nature by John Barrow
Includes chapters on naturalness, the anthropic principle, and dimensions.

Warped Passages by Lisa Randall
A speculative account of hidden dimensions.

Why String Theory? by Joseph Conlon
Not an easy read, quite technical, but accurately reflects the current state of the field.

PICTURE CREDITS

All photographs are public domain unless otherwise stated.

Thanks to Paul Noonan for the back cover photograph.

Photograph of dice is by Diacritica and is provided by Wikimedia Commons.

Photograph of ice skater Elena Sokolova is by K. "bird" N. and is provided by Wikimedia Commons.

Photograph of LIGO courtesy Caltech/MIT/LIGO Laboratory.

Diagram of the black hole merger was influenced by a diagram by Ben Gilliland in *Astronomy Now* magazine.

Calabi-Yau image is by Lunch and is provided by Wikimedia Commons.

Aeroplane axes image is by Auawise and is provided by Wikimedia Commons. Modifications made by Jrvz.

Joystick image is by Piotr Michal Jaworski and is provided by Wikimedia Commons.

HIDDEN
PLAINSIGHT 7

The fine-tuned universe

AVAILABLE NOW

159

Made in the USA
Lexington, KY
27 April 2017